Elements

ZINC, CADMIUM AND MERCURY

Zn Cd Hg

How to use this book

This book has been carefully developed to help you understand the chemistry of the elements. In it you will find a systematic and comprehensive coverage of the basic qualities of each element. Each two-page entry contains information at various levels of technical content and language, along with definitions of useful technical terms, as shown in the thumbnail diagram to the right. There is a comprehensive glossary of technical terms at the back of the book, along with an extensive index, key facts, an explanation of the Periodic Table, and a description of how to interpret chemical equations.

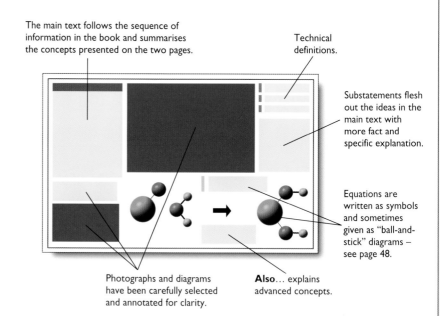

The main text follows the sequence of information in the book and summarises the concepts presented on the two pages.

Technical definitions.

Substatements flesh out the ideas in the main text with more fact and specific explanation.

Equations are written as symbols and sometimes given as "ball-and-stick" diagrams – see page 48.

Photographs and diagrams have been carefully selected and annotated for clarity.

Also... explains advanced concepts.

An Atlantic Europe Publishing Book

Author
Brian Knapp, BSc, PhD
Project consultant
Keith B. Walshaw, MA, BSc, DPhil
(Head of Chemistry, Leighton Park School)
Industrial consultant
Jack Brettle, BSc, PhD (Chief Research Scientist, Pilkington plc)
Art Director
Duncan McCrae, BSc
Editor
Elizabeth Walker, BA
Special photography
Ian Gledhill
Illustrations
David Woodroffe
Designed and produced by
EARTHSCAPE EDITIONS
Print consultants
Landmark Production Consultants Ltd
Reproduced by
Leo Reprographics
Printed and bound by
Paramount Printing Company Ltd

First published in 1996 by
Atlantic Europe Publishing Company Limited, Greys Court Farm,
Greys Court, Henley-on-Thames, Oxon, RG9 4PG, UK.

Suggested cataloguing location
Knapp, Brian
    Zinc, Cadmium and Mercury
    ISBN 1 869860 64 0
    – *Elements* series
540

Acknowledgements
The publishers would like to thank the following for their kind help and advice: *Jonathan Frankel* of *J. M. Frankel and Associates, Peter Johnson, Julie James, Rolls-Royce plc* and *British Petroleum International.*

Picture credits
All photographs are from the **Earthscape Editions** photolibrary except the following:
(c=centre t=top b=bottom l=left r=right)
Courtesy of **British Petroleum Company plc** 18br; courtesy of **Canon (UK) Ltd** 43cr; **The Hutchison Library** 37t (Richard House); **Mary Evans Picture Library** 38/39b; courtesy of **Rolls-Royce plc** 42; **Science Picture Library/US Department of Energy** 35tl; **UKAEA** 41bl and **ZEFA** 31b (D. Cattani), 39t (A. Ribeiro).

Front cover: Droplets of mercury being squeezed from a pipette. Mercury is the only metal that is liquid at room temperature.
Title page: Zinc reacts with dilute hydrochloric acid, releasing hydrogen gas.

This product is manufactured from sustainable managed forests. For every tree cut down at least one more is planted.

The demonstrations described or illustrated in this book are not for replication. The Publisher cannot accept any responsibility for any accidents or injuries that may result from conducting the experiments described or illustrated in this book.

Contents

Introduction

An element is a substance that cannot be broken down into a simpler substance by any known means. Each of the 92 naturally occurring elements is therefore one of the fundamental materials from which everything in the Universe is made. This book is about the elements zinc, mercury and cadmium.

Zinc

Zinc, symbol Zn, is the metal that makes the casing of most dry batteries, and coats the world's barbed wire fences. We usually see zinc as a dull grey metal, but when freshly polished it is bluish-white in colour. Zinc is very reactive and so is never found as native metal. Its reactivity makes it particularly useful because it readily attaches to other metals to make a zinc coating and easily mixes with other metals to make alloys such as brass.

Compared with many other metals, zinc has a low melting point (420°C), an advantage for metalworkers of the past whose furnaces could not reach the temperatures we can achieve today.

The ancient Egyptians were the first to use zinc, although they did so unintentionally! They made their brass from copper ores that were contaminated with zinc, and simply did not realise that they were using a combination of metals. By Roman times it was understood that zinc ore was needed to make brass, but nobody thought of any use for the metal on its own until the 18th century.

As soon as scientists understood that a reactive metal could be used to protect iron, they began to use zinc much more extensively. They also

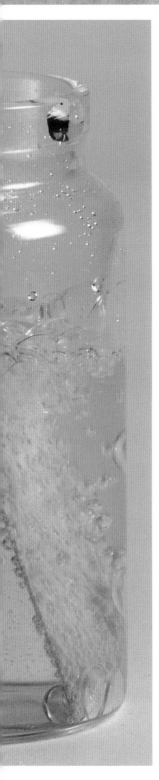

discovered a wide range of alloys in which zinc can play a vital role. About one-tenth of all zinc is now used for making the cases of dry cells (dry batteries).

Mercury

Mercury, symbol Hg, is the only metal that is a liquid at room temperature. It is one of the rarest metals, and yet it is found in many homes – in the form of the silvery liquid in thermometers.

Mercury is named after the planet and has been recognised since ancient times. This is perhaps because the bright red ore of mercury, called cinnabar, was attractive to ancient civilisations as a pigment for paint. Ancient civilisations also discovered that by simply heating the cinnabar, liquid mercury flows out.

The Greeks named the element *hydrargyros*, after its "quicksilver" properties. This is why the chemical symbol for mercury is Hg.

Cadmium

Cadmium, symbol Cd, gets its name from the old term for zinc ore, *cadmia*, because it was first discovered with zinc ore. Like zinc, cadmium is a reactive metal that readily plates onto other metals. Its special advantage is that cadmium has a very low friction, making it ideal for use on ball bearings and other moving parts. But it is in batteries and in photoelectric cells that cadmium becomes part of everyday life. Every time street lights come on automatically at dusk, or we rely on the automatic exposure meter of a camera, we are making use of photosensitive cells made with cadmium compounds.

◄ A piece of zinc reacting with dilute hydrochloric acid results in bubbles of hydrogen gas.

Zinc minerals

Zinc is found among many other metals in places where rocks were once molten. Zinc makes about two-thousandths of one per cent of the Earth's crust, making it the 25th most common element.

The most abundant and important zinc minerals are zinc sulphide (sphalerite or zinc blende) and zinc carbonate (smithsonite). Zinc sulphide is the most common of the two, occurring as dark crystals. It is a soft, easily scratched mineral. Geologists can identify it among similar-looking minerals (such as lead compounds) by rubbing it on the surface of an unglazed tile, because it will always leave a mark with a strong brown colour. Sphalerite is found in the veins that were formed above molten igneous rocks.

▼ At the top of a magma chamber, hot fluids push their way into cracks and fissures in the overlying rocks. Here the liquids cool, and the minerals solidify as crystals. These places, called hydrothermal veins by geologists, are where miners look for zinc ores, especially sphalerite.

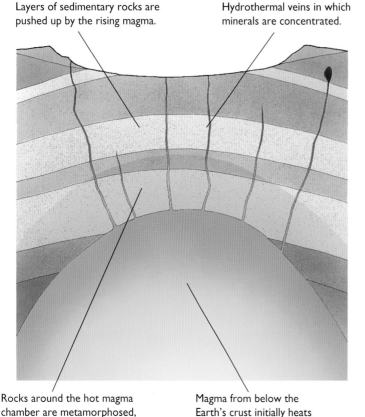

Layers of sedimentary rocks are pushed up by the rising magma.

Hydrothermal veins in which minerals are concentrated.

Rocks around the hot magma chamber are metamorphosed, or changed.

Magma from below the Earth's crust initially heats the surrounding rocks but eventually cools to form granite.

Smithsonite

Zinc carbonate, also called smithsonite, is found most commonly in limestone areas in veins that also contain lead. Zinc carbonate probably originated as a hot solution of zinc sulphide which was precipitated on reacting with the calcium carbonate (limestone).

Zinc carbonate is an important source of zinc for the mining industry. The name smithsonite comes from a 19th century English chemist who was the first to discover zinc carbonate in limestone. The prestigious Smithsonian scientific institution in Washington DC is also named for him, as he left a legacy in his will for its foundation.

crystal: a substance that has grown freely so that it can develop external faces. Compare with crystalline, where the atoms are not free to form individual crystals and amorphous where the atoms are arranged irregularly.

hydrothermal: a process in which hot water is involved. It is usually used in the context of rock formation because hot water and other fluids sent outwards from liquid magmas are important carriers of metals and the minerals that form gemstones.

magma: the molten rock that forms a balloon-shaped chamber in the rock below a volcano. It is fed by rock moving upwards from below the crust.

sulphide: a sulphur compound that contains no oxygen.

▼ The dark coloured crystalline material is sphalerite. The brassy cubes are pyrite (iron sulphide).

▶ Pieces of zinc.

▼ Zinc is best seen in its true bluish-white colour on materials where it has been newly applied, such as these galvanised steel nails. Over time the zinc reacts with oxygen in the air to form a coating of zinc oxide, which is dark grey in colour.

Extracting zinc from its ore

Zinc is a very reactive element and so is not found as a native metal. Because it is reactive, zinc is difficult to separate from its compounds, and as a result little use was made of the element before the 19th and 20th centuries.

The main use of zinc was as an accidental constituent of copper by the ancient Egyptians. It was only in 1721 that German metallurgist Johann Henckel managed to extract zinc as a metal. Nevertheless, it was a century before it came into more widespread use because a commercial method of smelting the ore could not be found.

In fact, in many cases even today, the cost of smelting zinc alone from the low grade ore in which it is found would make it uneconomic. However, it is usually associated with other elements such as lead, copper, gold and silver. Thus, by extracting a number of elements, zinc can be recovered economically.

As with other low-grade ores, the metal has to be extracted in a number of steps, designed to concentrate the ore and remove as much waste rock as possible before the expensive smelting process begins.

When the ore has been enriched, it can be smelted; but to achieve the highest quality, a further stage of refining is necessary using an electrolytic cell.

Processing zinc ore

The first stage of zinc processing is called froth flotation. The ore is crushed to a fine powder and placed into a vat with water, and wetting and frothing chemicals.

Jets of air are introduced into the vat, causing the frothing agents to make bubbles. The chemicals in the vat prevent the metals from becoming wetted by the water. Because they remain dry they can be caught up by air bubbles and lifted to the surface of the vat, where they can be floated off and collected. The wet particles of rock, by contrast, sink to the bottom.

After flotation, the ore may be concentrated to produce up to one-half metal. At this concentration it can be roasted, which changes the zinc sulphide into zinc oxide. A byproduct of the roasting operation, sulphur dioxide, is usually converted to sulphuric acid and sold.

Zinc oxide is then smelted with a supply of coke in a furnace. The coke, being almost pure carbon, reduces the oxide to metal and produces carbon monoxide gas. The zinc metal, which has a low boiling point, forms a vapour in the smelter, and this is distilled and collected in order to make it into ingots. The overall reaction is shown opposite.

electrode: a conductor that forms one terminal of a cell.

electrolysis: an electrical–chemical process that uses an electric current to cause the break up of a compound and the movement of metal ions in a solution. The process happens in many natural situations (as for example in rusting) and is also commonly used in industry for purifying (refining) metals or for plating metal objects with a fine, even metal coating.

electrolyte: a solution that conducts electricity.

ore: a rock containing enough of a useful substance to make mining it worthwhile.

oxide: a compound that includes oxygen and one other element.

pyrometallurgy: refining a metal from its ore using heat. A blast furnace or smelter is the main equipment used.

sulphide: a sulphur compound that contains no oxygen.

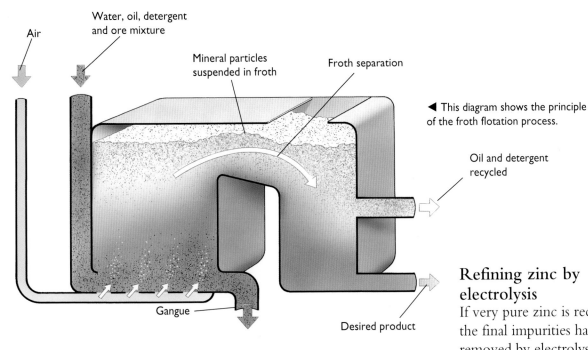

◀ This diagram shows the principle of the froth flotation process.

Refining zinc by electrolysis

If very pure zinc is required, the final impurities have to be removed by electrolysis. In this process, an electric current is passed between two electrodes in a solution that conducts electricity. It is widely used to refine metals to high purity.

The smelted zinc is dissolved in sulphuric acid to form zinc sulphate. The zinc sulphate solution is used as the electrolyte and zinc metal is collected at the aluminium or stainless steel cathode. Zinc-plated sheets produced in this way are separated from their electrodes by being melted in a furnace and then cast into ingots.

EQUATION: Removal of sulphur in zinc ore by heating

Zinc sulphide + zinc oxide ⇨ zinc + sulphur dioxide

$$ZnS(s) \quad + \quad 2ZnO(s) \quad ⇨ \quad 3Zn(s) \quad + \quad SO_2(g)$$

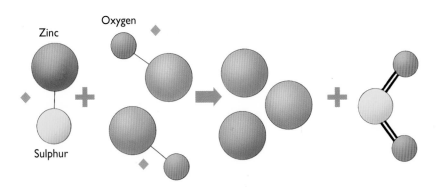

The reactivity of zinc

Zinc is amphoteric, that is, it is soluble in alkalis and acids but insoluble in water. However, many of its compounds are soluble in water. Most zinc compounds are white.

Few other common elements are more reactive than zinc. This, above all, makes zinc useful for protecting other metals, such as steel. If steel and zinc are exposed to water (which contains dissolved oxygen and carbon dioxide gas) the zinc corrodes to zinc carbonate, while the steel remains unchanged.

❶▲ A zinc strip is laid across a dish containing copper sulphate solution.

Reaction of zinc and copper

When a piece of zinc strip is dipped into a solution of copper sulphate, some of the zinc begins to react with it. Quite soon some of the copper in the solution can be seen to coat the zinc and the blue solution turns paler. Over time the zinc strip dissolves away and the solution becomes completely colourless.

The copper and the zinc have reacted so that the zinc atoms have lost electrons to the copper. As this happens the zinc becomes soluble and the copper, which now has the same number of electrons as protons (i.e. it forms atoms), becomes insoluble and is deposited on the zinc or as a fine powder at the bottom of the dish.

Copper metal is precipitated.

❷▲ The copper is precipitated on the zinc, while some of the zinc goes into solution.

EQUATION: Displacement of copper by zinc

Zinc + copper sulphate ⇨ zinc sulphate + copper

$$Zn(s) + CuSO_4(aq) \Rightarrow ZnSO_4(aq) + Cu(s)$$

REACTIVITY SERIES	
Element	Reactivity
potassium sodium calcium magnesium aluminium manganese chromium **zinc** iron cadmium tin lead **copper** mercury silver gold platinum	*most reactive* *least reactive*

▲ The reactivity series of metals shows that zinc is more reactive than copper.

acid: compounds containing hydrogen which can attack and dissolve many substances. Acids are described as weak or strong, dilute or concentrated, mineral or organic.

amphoteric: a metal that will react with both acids and alkalis.

electrolyte: a solution that conducts electricity.

▲ A piece of zinc reacting with dilute hydrochloric acid. The bubbles are hydrogen gas.

Also...

The reaction of zinc with a copper salt occurs because zinc is more reactive than copper. The transfer of electrons from one metal to another is the basis of all primary batteries (see page 12). Luigi Galvani was the first person to discover that when two metals are connected through certain types of fluid (electrolytes), then an electric current will flow. Galvani used iron and brass rods and the body fluids of frogs in his famous experiments of the 19th century (in which he made frogs' legs twitch). Today zinc is commonly used as an electrode material in dry batteries, but all cells of this kind are called galvanic cells in honour of Galvani.

EQUATION: The reaction of zinc with an acid

Dilute hydrochloric acid + zinc ⇨ zinc chloride + hydrogen gas

$$2HCl(aq) \quad + \quad Zn(s) \quad ⇨ \quad ZnCl_2(aq) \quad + \quad H_2(g)$$

Chlorine

Hydrogen

Zinc

Zinc cells

On pages 10 and 11 we saw how zinc reacted with a copper salt solution. The flow of ions involved in the exchange of metals is actually a flow of electricity, but this is not noticeable because the ions have not been channelled into an orderly flow (such as happens, for example, with the flow of electrons in a wire). The first person to make a cell that produced electricity was Alessandro Volta in 1796. He saw that if the zinc could be kept separate from the copper, then he could make ions flow in an orderly way within the cell. In this way, he could create an electric current (of electrons) in a wire outside the solutions. Using these principles, he created a chemical battery using an electrolyte of animal hide soaked in vinegar.

The difficulty with the Volta cell was that one of the electrodes suffered badly from corrosion. To overcome this problem (and to make it possible to use a battery to supply a steady flow of electricity for the newly invented telegraph), in 1836 the British chemist John Daniell made a chemical battery with zinc, in a porous pot of zinc sulphate and copper, in a flask of copper sulphate. When a wire was connected between them, ions could flow through the porous pot, thus producing a flow of electricity.

Further development of the Daniell cell led to the Leclanché cell, or common dry battery (shown on page 14).

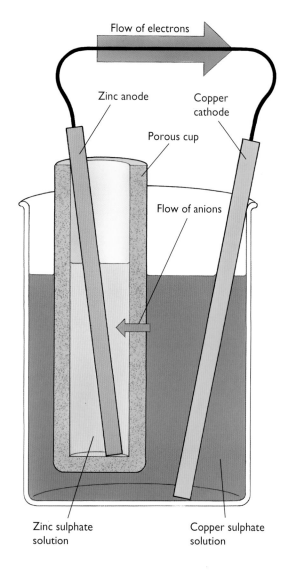

▼ A Daniell cell. Notice that the current in the electrolyte is carried by ions, whereas the current in the wire of the external circuit is carried by electrons.

Flow of electrons

Zinc anode

Copper cathode

Porous cup

Flow of anions

Zinc sulphate solution

Copper sulphate solution

Also...

The natural voltage of a Daniell cell using zinc and copper is 1.1 volts. The voltage of a dry cell (page 14) is 1.5 volts. Other combinations of elements produce their own unique natural voltages.

Secondary batteries made with zinc

A secondary battery can be charged from an external source, causing a chemical change of the surface of the electrodes. When the battery is connected to a load, the chemical change reverses and electricity is given out.

The most commonly known secondary battery, used in cars, is the lead-acid battery. However, zinc–silver batteries are also used as secondary batteries. They are much more expensive than lead-acid batteries, but will produce a high current very quickly and store about three times as much electrical energy per gram of battery compared with lead-acid batteries. Their main use is for military purposes.

Cadmium–silver batteries are used in the same way as zinc–silver batteries. Neither is able to store a charge for a long time, and chemical reaction occurs on the plates, resulting in a loss of charge. Zinc–cadmium batteries discharge at a high rate on demand, but they are also able to store a charge for long periods.

The electrolyte used in these secondary batteries is potassium hydroxide. Each zinc–silver cell produces 1.6 or 1.8 volts (depending on the oxide of silver used). Dry zinc–silver batteries (where the electrolyte is made into a paste) are sometimes used in hearing aids.

anode: the negative terminal of a battery or the positive electrode of an electrolysis cell.

cathode: the positive terminal of a battery or the negative electrode of an electrolysis cell.

cell: a vessel containing two electrodes and an electrolyte that can act as an electrical conductor.

electrode: a conductor that forms one terminal of a cell.

electrolyte: a solution that conducts electricity.

electron: a tiny, negatively charged particle that is part of an atom. The flow of electrons through a solid material such as a wire produces an electric current.

ion: an atom, or group of atoms, that has gained or lost one or more electrons and so developed an electrical charge. Ions behave differently from electrically neutral atoms and molecules. They can move in an electric field, and they can also bind strongly to solvent molecules such as water. Positively charged ions are called cations; negatively charged ions are called anions. Ions carry electrical current through solutions.

How a galvanic cell works

Galvanic cells, of which the Daniell cell is a special example, are made of a container containing two compartments, allowing each electrode to be bathed in its own electrolyte. The compartments are connected by a salt bridge, an inverted U-shaped tube containing a jellified concentrated salt (usually potassium chloride). This allows ions to flow between the electrodes and complete the circuit. As we see on pages 14 and 15, in practical cells, such as "dry cells", the salt bridge is reshaped and sandwiched between two electrodes.

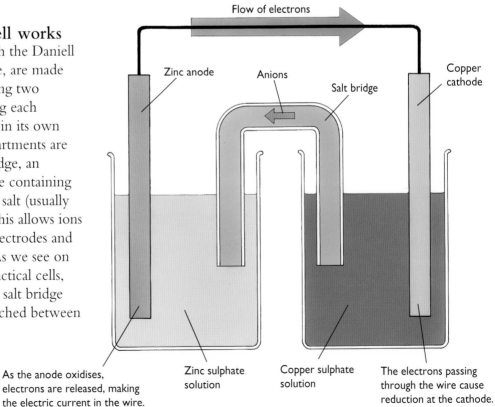

Flow of electrons

Zinc anode   Anions   Salt bridge   Copper cathode

As the anode oxidises, electrons are released, making the electric current in the wire.

Zinc sulphate solution

Copper sulphate solution

The electrons passing through the wire cause reduction at the cathode.

Dry cells

The dry cell is a portable form of chemical cell in which the zinc acts as the case as well as an electrode. To prevent spillage of liquids, the solutions of the Daniell cell (see page 12) are made into pastes. The dry cell was first made by Georges Leclanché.

The problem with this arrangement is that the zinc is gradually used up in the reaction, so holes develop in the case and the cell begins to leak. This is the reason newer "leakproof" cells were developed.

Alkaline batteries are also Leclanché cells, but with a different electrolyte.

All Leclanché-type batteries have a zinc anode (negative terminal), which also acts as a container, and a carbon and manganese dioxide cathode (positive terminal). The electrolyte is ammonium chloride. The terminals are separated by a porous material that can be as simple as a sheet of paper.

During use, oxidation takes place at the zinc anode, thereby corroding the zinc case. Manufacturers design the case to be sufficiently thick so that it does not corrode right through before the electrical energy has been used up.

▼ A sequence of diagrams to show how the galvanic type of cell shown on page 13 can be modified to make the everyday dry cell. Notice that the copper electrode has been replaced with carbon, zinc forms the case, and the salt bridge has become a sheet of lining paper. By such modifications the dry cell can be made cheaply. Billions of dry cells are now made each year.

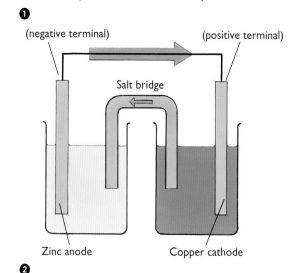

❶
(negative terminal) (positive terminal)
Salt bridge
Zinc anode   Copper cathode

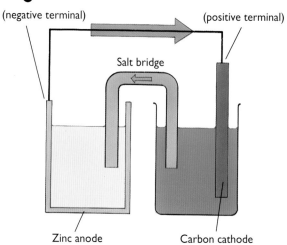

❷
(negative terminal) (positive terminal)
Salt bridge
Zinc anode   Carbon cathode

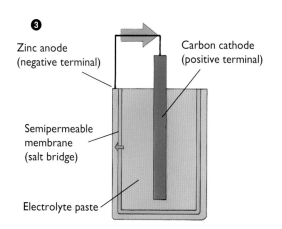

❸
Zinc anode (negative terminal)   Carbon cathode (positive terminal)
Semipermeable membrane (salt bridge)
Electrolyte paste

Also...

The zinc dry cell is one of a number of primary dry batteries that use zinc. In addition to the inexpensive zinc cell, alkaline, mercury and silver-containing zinc batteries are also made.

The only difference between the alkaline cell and the zinc cell is that potassium hydroxide is used as the electrolyte. With this modification, the power output can be kept high for a larger part of the cell's life than with an ordinary ammonium chloride electrolyte.

Mercury dry cells are not as common nowadays because of concern over mercury poisoning. Mercury cells are essentially alkaline cells that use mercury oxide as the central terminal. Like other alkaline cells, mercury cells will maintain a steady voltage over more of their working life than ordinary cells.

Cells made with silver chloride or silver oxide are also popular for special applications such as hearing aid batteries. Silver chloride is used as the positive terminal and zinc as the negative terminal, with zinc, sodium or ammonium chloride as an electrolyte.

electrode: a conductor that forms one terminal of a cell (battery).

electrolyte: a solution that conducts electricity.

oxidation: a reaction in which the oxidising agent removes electrons. (Note that oxidising agents do not have to contain oxygen.)

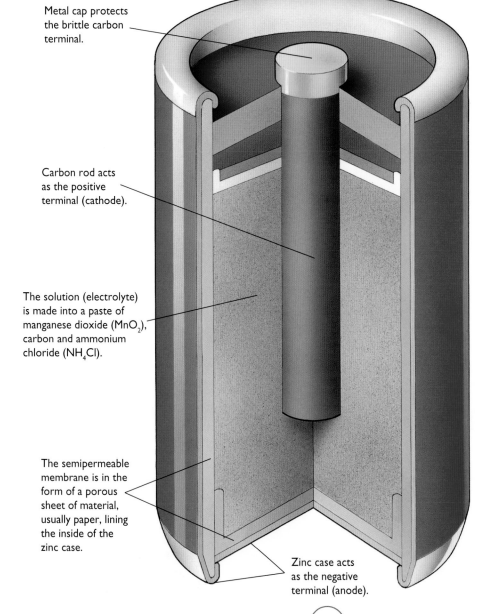

Metal cap protects the brittle carbon terminal.

Carbon rod acts as the positive terminal (cathode).

The solution (electrolyte) is made into a paste of manganese dioxide (MnO_2), carbon and ammonium chloride (NH_4Cl).

The semipermeable membrane is in the form of a porous sheet of material, usually paper, lining the inside of the zinc case.

Zinc case acts as the negative terminal (anode).

▲▼ The zinc casing of zinc batteries is the anode of the cell and therefore certain to corrode, as shown in the battery above. Eventually this will allow the electrolyte to leak out. The one below has a protective leakproof casing.

Galvanising

Galvanising takes its name from Luigi Galvani, one of the pioneers of electrochemistry. Galvanising is the application of a thin layer of zinc over the entire surface of another metal, such as steel, in order to protect it from corrosion.

Zinc is a naturally reactive material and quickly reacts with air to produce an invisible gastight surface of zinc oxide. The advantage of zinc over other plating metals such as tin, is that because it is more reactive than steel, zinc will still protect the steel even if the zinc is partly scraped away, as might happen if the surface were deeply scratched.

The reactivity of metals
Some metals, such as iron, are subject to corrosion when placed in air or soil that is damp.

By connecting metals together in a damp environment, a natural cell is formed. In a cell, one of the electrodes (the anode) always corrodes, while the other (the cathode) remains undamaged.

Metals higher up in the reactivity series (see page 11) will always protect those below them, with the more reactive metal behaving as the anode. Thus zinc will protect iron because it is above iron in the reactivity series.

When the surface of zinc-coated (galvanised) iron is scratched, the iron could give up electrons to the oxygen in the air; but they are replaced by electrons released from the zinc. Thus, the iron suffers no change itself and does not rust, while the zinc is gradually used up.

▼ The main cables of the Brooklyn Bridge are made of bundles of galvanised cast steel wires nearly five millimetres thick. It was the world's first use of galvanised wire for large structures.

electroplating: depositing a thin layer of a metal onto the surface of another substance using electrolysis.

flux: a material used to make it easier for a liquid to flow. A flux dissolves metal oxides and so prevents a metal from oxidising while being heated.

galvanising: applying a thin zinc coating to protect another metal.

◀ Galvanised buckets being used to carry latex in a rubber plantation. It is important that the latex does not get contaminated by rust. The galvanised surface shows a crystalline pattern.

Galvanising methods

Galvanising is done mainly by dipping metals in a bath of molten zinc. Before dipping, the metal is degreased using a solvent and any surface oxide is removed by dipping it in an acid (known as pickling). Then the surface has a flux coated on, in this case a solution of zinc chloride which leaves the metal in a condition to take an even coating of zinc.

Most zinc is plated at a temperature of about 850°C. The coat will only stick if the metal being coated reaches the temperature of the zinc bath, so the metal must be immersed for a while. During the galvanising process the zinc partly amalgamates with the metal it is coating and only the outer surface of the coat is pure zinc.

Objects that cannot be left in a hot bath are galvanised using more expensive electroplating techniques. The object to be plated is made into the cathode of an electrolytic cell, while the anode (the source of the zinc) is made from a zinc sheet. An electric current is then passed through the cell.

Galvanising produces a coating of zinc that is between 2 and 15 microns thick, different thicknesses being used for different purposes.

◀ Galvanised steel roofs are one of the world's most common roofing materials. The one in the foreground of this picture has been painted the colour of tiles to make it look more attractive.

Zinc protects large structures

Objects such as ships, pipelines, oil or natural gas storage tanks and bridges are too large to be protected by galvanising. Instead, they are protected either using paint or by connecting them to a number of large zinc blocks buried in damp ground. The moisture in the damp ground provides the electrolyte and allows the object and attached zinc block to behave as a battery. The zinc blocks supply ions to the steel, becoming corroded away in the process. The steel acquires the ions and releases electrons to the air. As a result, the iron is not corroded.

The zinc blocks are called sacrificial anodes because they are the anodes of the cell that forms between the damp environment, steel structure and protective zinc block. The blocks are sacrificed to protect the steel structure, it being a relatively easy process to replace the zinc periodically.

anode: the negative terminal of a battery or the positive electrode of an electrolysis cell.

cathodic protection: the technique of making the object that is to be protected from corrosion into the cathode of a cell. For example, a material, such as steel, is protected by coupling it with a more reactive metal, such as magnesium. Steel forms the cathode and magnesium the anode. Zinc protects steel in the same way.

electrode: a conductor that forms one terminal of a cell.

electrolysis: an electrical–chemical process that uses an electric current to cause the break up of a compound and the movement of metal ions in a solution. The process happens in many natural situations (as for example in rusting) and is also commonly used in industry for purifying (refining) metals or for plating metal objects with a fine, even metal coating.

electrolyte: a solution that conducts electricity.

◄ **Zinc-based paints**
One increasingly important use of zinc is in the form of high purity powder called zinc dust. This can then be incorporated in corrosion-resistant paints and other coatings. This picture shows a zinc-painted section of the Sydney Harbour Bridge, Australia.

◄▼ The photograph on the left shows a section of the trans-Alaska pipeline being installed. To protect the steel pipe underground, twin ribbons of zinc are wrapped around it in a corkscrew pattern before the pipe is buried. The pipe and zinc ribbons are further protected by encapsulation in a plastic coating.

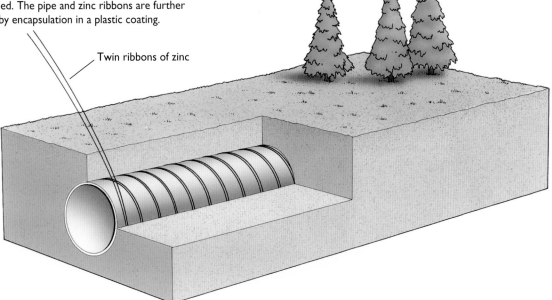

Twin ribbons of zinc

Brass and bronze

When zinc is melted, it readily combines with a number of other metals to make metal mixtures, or alloys Of these, the most important is brass, an alloy of zinc and copper with traces of other elements. In general, unless the proportion of zinc exceeds one-third of the alloy, the greater the proportion of zinc, the stronger the brass.

Many types of brass are produced, each varying in the proportion of copper to zinc. The kinds of brass that are most easy to draw out into tubes or wires have relatively little zinc in them, perhaps one-fifth of the total alloy. The stronger brasses, with greater zinc content, are used for keys and other parts where strength and machinability are needed, and where softness would therefore be a major handicap. The strongest brass of all has one-third zinc along with 2% lead. It is known as leaded brass.

Bronze is an alloy of copper significantly different from brass. Bronze is a copper alloy with tin as its major secondary constituent. Zinc is used to give the bronze important special properties, such as ease of casting.

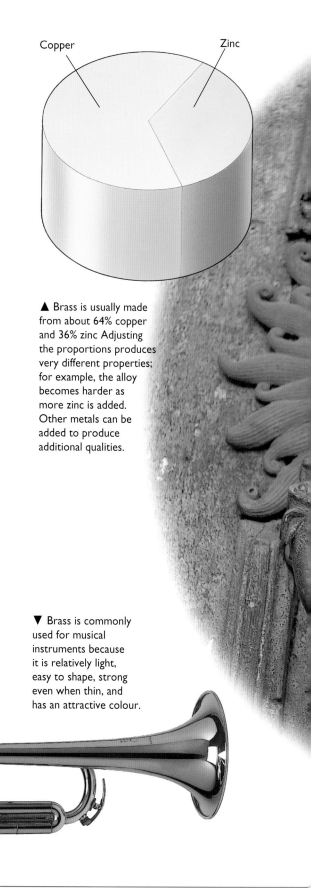

Copper   Zinc

▲ Brass is usually made from about 64% copper and 36% zinc Adjusting the proportions produces very different properties; for example, the alloy becomes harder as more zinc is added. Other metals can be added to produce additional qualities.

▼ Brass is commonly used for musical instruments because it is relatively light, easy to shape, strong even when thin, and has an attractive colour.

▼ The first brass was made by the Romans about two thousand years ago. They found it especially valuable for making coins. Brass was also popular because it has an attractive colour and it is much cheaper and stronger than pure copper. This is a brass "sanctuary" knocker on the door of a cathedral.

alloy: a mixture of a metal and various other elements.

corrosion: the *slow* decay of a substance resulting from contact with gases and liquids in the environment. The term is often applied to metals. Rust is the corrosion of iron.

▶ Brass is cast (as in taps), machined (as in screw threads on plumbing connectors), or punched out and then machined, as in cylinder lock keys.

The cylinder lock

The cylinder or pin-tumbler lock, often popularly called the "Yale" lock after its American inventor, is made out of brass for a number of reasons. For example, brass does not corrode readily, a major advantage when using locks fitted to street doors that are exposed to the weather.

The pins, the cylinder and all the other components can be cast and turned easily, yet they are strong enough to resist attempts to break the lock.

▶ Bronze is usually made from about 78% copper and 12% tin. Adding zinc and lead to the bronze alloy produces a material that is much more suitable for casting.

An alloy with about nine-tenths copper and equal proportions of the other metals is called gunmetal. It was commonly used in cannon, not only for its corrosion resistance but also for its machinability.

Casting using zinc alloys

Casting, pouring liquid metal into a mould and allowing it to solidify, is a process used to make complicated shaped objects. It is one of the oldest metallurgical techniques and is used to produce many of the bronze and brass objects described in this book. The shapes produced by casting are also used in many machines.

Two methods are used in casting. In the traditional method a mould is made to the required shape. It may be made of sand if the shape is not to be too precise, or it may be cut from another metal or some other hard material that will not melt when the liquid metal is poured.

For most precision parts a different system is used. The mould is first made of metal or some other material that will stand up to high pressure. The molten metal is then injected into the mould.

The crucial property of the metal being used for die-casting is that it should flow easily. A zinc alloy has just these properties. A die-casting alloy is typically 96% zinc, 4% aluminium and trace amounts of copper and magnesium. The addition of these other metals makes the alloy as strong as steel. The addition of copper is crucial in ensuring that the casting doesn't warp as it sets.

The use of zinc in castings now consumes as much zinc as is used in galvanising.

▲ The body of this vehicle component is cast in a zinc-based alloy. Notice the rough texture to the surface due to the casting.

▼ **Die-cast models**
Most metal model trains, vehicles and so on are made with a zinc-based alloy using the die-cast process. This gives a high precision toy whose metals have no harmful effects. (In the past lead alloys were also used, but lead compounds have subsequently been discovered to be potentially toxic.)

Die-cast objects are also widely found in business machines and electrical appliances.

▼ This sequence of diagrams shows the steps involved in preparing for casting.

alloy: a mixture of a metal and various other elements.

❶▶ A "former" is made of wood or another hard but easily shaped material. This is placed in a box of tightly packed sand. A resin is added to improve the moulding properties of the sand and to ensure the sand sets hard. The two halves of the box are clamped together.

These shapes, called "nibs", are added to provide paths for the flow of metal and escape of air.

❷▼ The two halves of the containing box are separated, and the "former" removed to leave the mould.

❸▼ The box halves are clamped together again. Molten metal is then poured into the mould and allowed to cool and form a casting.

❹▼ The mould is opened, the casting removed and the "nibs" cut or broken off.

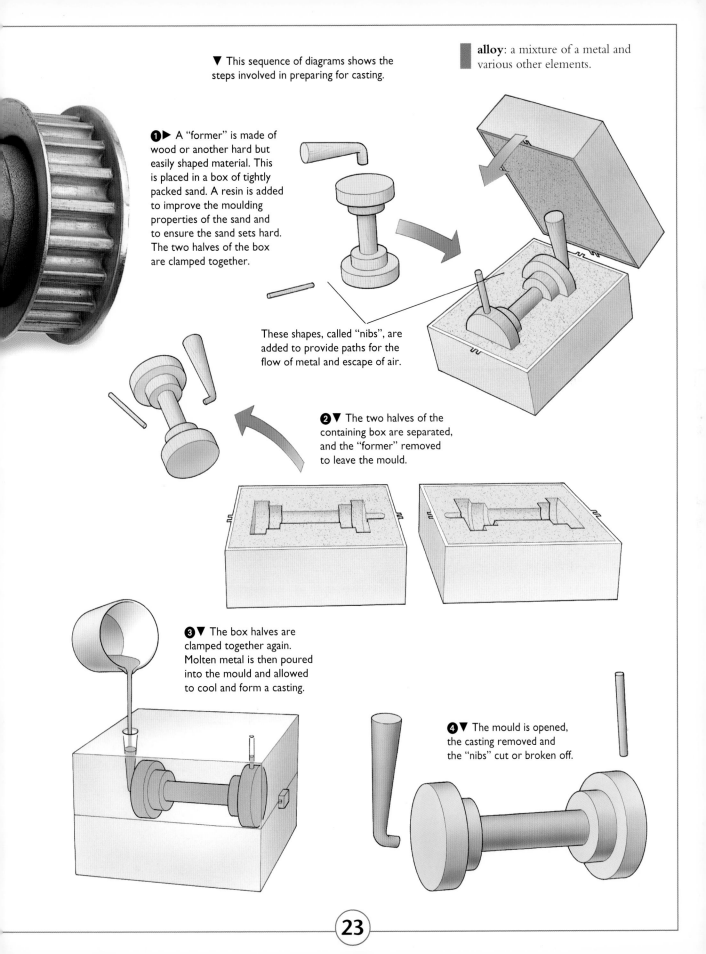

Compounds of zinc

Most zinc is used in alloys, but zinc compounds also have widespread applications, from pesticides and fungicides to strengtheners in rubber. The special reflective properties of zinc oxide also make it extremely useful as harmless paint pigments (instead of lead).

Zinc oxide

Zinc oxide (ZnO) is a common, white compound of zinc. It is prepared by heating zinc in a furnace and then oxidising and finally distilling the vapour.

Zinc oxide has a high refractive index, meaning that light is reflected from it easily. It is therefore used to provide the colour (pigment) to white paint and to make a white glaze on ceramics. It is also used as one of the components for protecting skin from sunburn.

Zinc oxide is extremely toxic to mildew and fungi, but harmless to people, so it can be used widely in the home, for example in paints, in wallpaper adhesives and as the foundation for many forms of make-up.

Zinc oxide is also used in rubber. By adding between 3 and 5% zinc oxide to the rubber, curing is made much easier. When used in tyres, it helps to conduct heat away from the rubber so that the tyres do not get too hot. It also increases the strength of the rubber.

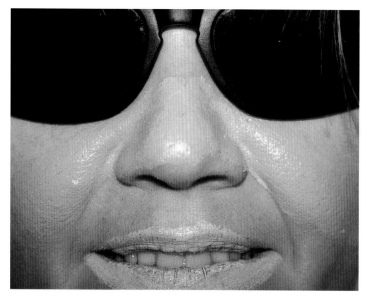

◀▼ Zinc oxide is used as a sunblock because it is highly reflective. The sunblock shown in the pot and on this skier's face has been coloured for cosmetic purposes; the sunblock used on the skier's lips has no added colours.

Zinc sulphide

Zinc sulphide is deposited on the inside of fluorescent tubes (see also uses of mercury, page 35) and cathode ray tubes (for televisions and computers). When bombarded by electrons, the particles of zinc sulphide glow, producing the patterns of light that we interpret as images.

Zinc chloride

Zinc chloride ($ZnCl_2$) is used as a flux. It is also used as a wood preservative and in batteries (see page 15).

Zinc sulphate

Zinc sulphate ($ZnSO_4$) is used in the manufacture of rayon. It also makes an important trace fertiliser and is helpful as a fungicide and pesticide, especially in citrus groves.

Zinc sulphate is also one of the chemicals in the frothing agent used in the flotation process for enriching metals (see page 8).

▲ Zinc oxide is a common additive to white exterior and gloss paints.

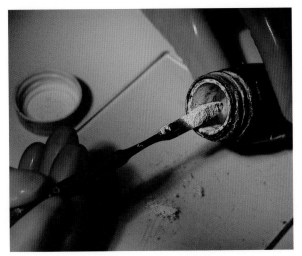

◄ Zinc oxide powder being used in a dental formulation.

Mercury

Mercury, or quicksilver, is a shiny liquid metal, thirteen and a half times as heavy as an equal volume of water. It is so dense that even lead can float on the surface of mercury. The name quicksilver, or living silver, came about from the silvery globules that form and roll about when mercury is poured onto a surface.

Mercury is a rare metal, yet it was one of the first metals ever found. Its ore, called cinnabar, is bright red and was used by the Chinese for three thousand years as a colour for paint. Cinnabar is also found in ancient Egyptian tombs dating back nearly 4000 years

The ancient Greeks used mercury as a medicine. In fact, mercury compounds can be used as disinfectants, but mercury itself is very toxic and can do far more harm than good! Mercury is poisonous because it can "switch off" enzymes in the body.

Mercury is a liquid between -39°C (at which it solidifies) and 35°C (at which it boils). It has a high surface tension, which is why it forms globules rather than spreading out over surfaces.

Alloys of mercury with another metal are called amalgams. Mercury makes liquid alloys with gold, silver, copper and lead.

Mercury gives off a toxic vapour when exposed to the air, even at temperatures well below its boiling point. This limits its uses in many applications. The main uses for mercury are as amalgams and in thermometers and barometers.

▲ Mercury's strong surface tension keeps the liquid formed as small globules.

Cinnabar

Mercury makes 83 parts per billion of the world's surface rocks, just a little more plentiful than gold. It is not thought of as a rare metal, however, because it is found in naturally concentrated form. Its main ore is mercury sulphide (HgS), called cinnabar.

To refine the ore the rock is ground up and heated, causing mercury vapour to rise from the ore. The vapour is distilled.

Cinnabar is found in places where there has been volcanic activity, where it was probably precipitated from hot waters rising above a magma chamber. The main mine for cinnabar is at Almaden, Spain. This mine has been worked since Roman times.

amalgam: a liquid alloy of mercury with another metal.

distillation: the process of separating mixtures by condensing the vapours through cooling.

enzyme: organic catalysts in the form of proteins in the body that speed up chemical reactions. Every living cell contains hundreds of enzymes, which ensure that the processes of life continue. Should enzymes be made inoperative, such as through mercury poisoning, then death follows.

ore: a rock containing enough of a useful substance to make mining it worthwhile.

surface tension: the force that operates on the surface of a liquid, which makes it act as though it were covered with an invisible elastic film.

vapour: the gaseous form of a substance that is normally a liquid. For example, water vapour is the gaseous form of liquid water.

Mercury vapour

If mercury is heated in a glass tube, the liquid will give off vapour. The vapour will condense back to a liquid in the upper, cooler regions of the tube, producing a mercury mirror.

The fact that mercury will vaporise readily is extremely important for metal refining. An ore is first mixed with mercury to make an amalgam (see page 36). This separates the metal from its ore. If the amalgam is heated, the mercury vaporises, leaving the desired metal behind. Gold prospectors use mercury extensively.

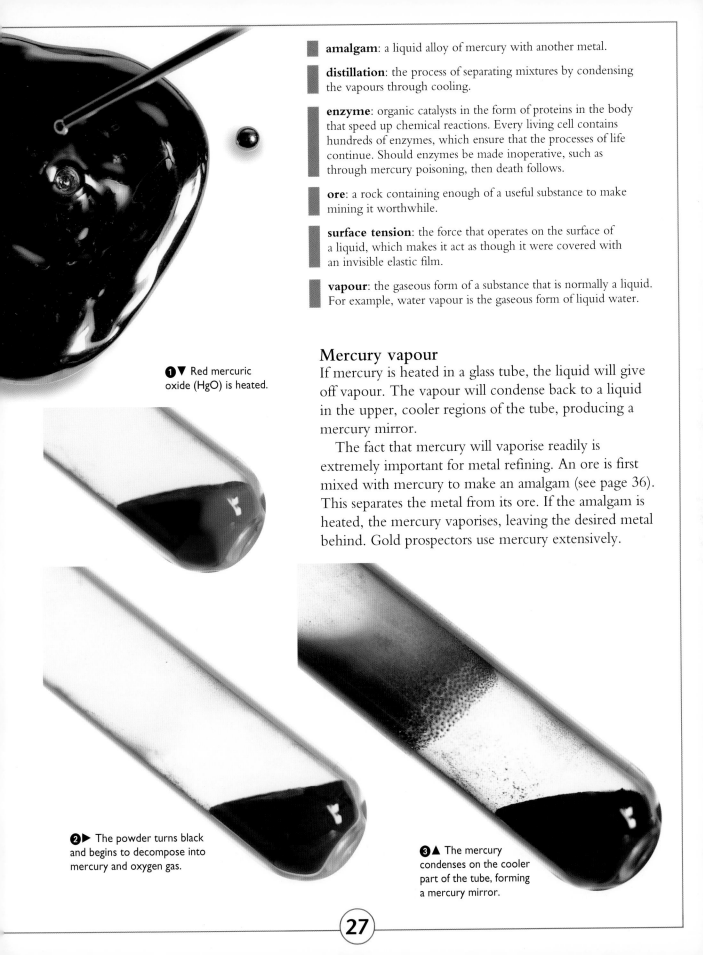

❶▼ Red mercuric oxide (HgO) is heated.

❷▶ The powder turns black and begins to decompose into mercury and oxygen gas.

❸▲ The mercury condenses on the cooler part of the tube, forming a mercury mirror.

Reactions with mercury

Mercury is one of the noble metals and thus is relatively unreactive. It will not react with either hydrochloric or sulphuric acid. However, fuming nitric acid, which is an extremely corrosive oxidising agent, will react violently with mercury.

Mercuric compounds will react more readily than pure mercury, as shown in the example at the top of page 29.

❶▼ Fuming nitric acid is added to a small amount of mercury (the shiny metal at the bottom of the beaker). This reaction produces mercuric nitrate and releases brown nitrogen dioxide.

❷▲ The violent reaction occurs. Notice that this is an exothermic reaction, giving out considerable amounts of heat.

EQUATION: Mercury and nitric acid

Mercury + nitric acid ⇨ mercuric nitrate + nitrogen dioxide + water

$$Hg(l) + 4HNO_3(l) \Rightarrow Hg(NO_3)_2(aq) + 2NO_2(g) + 2H_2O(l)$$

exothermic reaction: a reaction that gives heat to the surroundings. Many oxidation reactions, for example, give out heat.

noble metal: silver, gold, platinum, and mercury. These are the least reactive metals.

oxidising agent: a substance that removes electrons from another substance (and therefore is itself reduced).

◄ When a solution of yellow potassium chromate is dropped into a colourless solution of mercuric nitrate, an orange precipitate of mercuric chromate forms.

❸► When water is added, the mercuric nitrate is dissolved. As the reaction is completed, a colourless solution of mercuric nitrate remains.

(Notice that there was an excess of mercury used in this demonstration, which remains unchanged at the bottom of the beaker.)

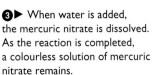

Using liquid mercury

Mercury is the only metal that is liquid at room temperatures.
It has a high coefficient of expansion, it is very dense, and it is a good
conductor of electricity. Each of these properties has been exploited
in applications from thermometers and barometers to electrical switches.

The mercury barometer

The fact that air has weight
was discovered by Evangelista
Torricelli in 1643. Using a tall
tube that had been filled with
mercury then inverted into a
bath of mercury, he found that
the mercury did not run out,
but that a vacuum was created
in the top of the tube and that
the mercury was supported by
air pressure.

He noticed that the height
of mercury in the tube changed
with weather conditions, thus
behaving as a measure of air
pressure. This was the foundation
of the barometer.

All liquids can be supported
in tubes; but since other liquids
are less dense than mercury, the
tubes have to be much taller
before a vacuum space is created
at the top. A tube using water,
for example, would have to be
10 m high, whereas a tube using
mercury only has to be 760 mm
high. This makes a mercury-filled
barometer a practical instrument
to use. A mercury barometer is
also known as a Fortin barometer.

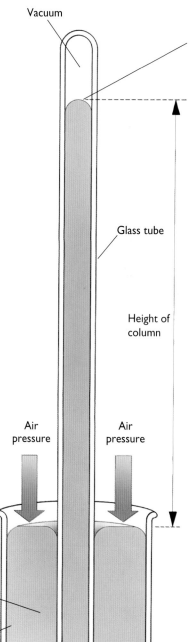

Vacuum

The upper curved surface of the
mercury is called a meniscus. It is the
opposite of a water meniscus, which
curves down rather than up.

Glass tube

Height of
column

Air
pressure

Air
pressure

Mercury

Container

▶ A diagram of a
mercury barometer.

Mercury switches

Mercury is a good conductor
of electricity and can be used
as a switch.

A mercury switch is a closed
tube containing two electrodes
and a small amount of mercury.
When the mercury runs between
the contacts, an electrical
connection is made; if it flows
away from the contacts, the
connection is broken.

This property can be used in
two ways: for detecting movement
and for detecting changes in
temperature. In a tilting switch,
as the tube tilts, the mercury flows
away from one of the contacts
and the electrical connection
is broken. If the switch moves
upright, the mercury flows back
and contact is made once more.

In a thermostat, as the
temperature rises, the mercury
expands until it touches the
contacts and makes a connection.
When it cools, it contracts from
the contacts and the connection
is broken.

◄ A mercury thermometer.

The mercury thermometer

All metals expand and contract considerably as the temperature rises and falls. Mercury has the added advantage of being a liquid. This means it can be placed in a narrow tube.

By confining the mercury to a narrow (small-bore) tube, the changes in volume of the mercury can be made to show up more easily.

Other liquids expand and contract with changing temperature. Water and alcohol could both be used in thermometers. However, mercury has a wider range of use: it does not freeze until the temperature falls to nearly –40°C, and it does not boil until 357°C. By contrast, water freezes at 0°C and boils at 100°C.

▼ Towards the end of the 19th century, mercury baths were used to float the lamp housing in lighthouses.

meniscus: the curved surface of a liquid that forms when it rises in a small bore, or capillary tube. The meniscus is convex (bulges upwards) for mercury and is concave (sags downwards) for water.

Lighthouse lamp bearings

Mercury has been used widely inside lighthouses since the end of the 19th century. Before this time lights were fixed. But by floating the lights in a tank of mercury, the lights could be made to rotate quickly with almost no friction and using just a small motor.

The process of floating lights on mercury led to a new way of using lighthouses. Rotating beams could be made to shine a unique pattern of light. This enabled sailors to identify each lighthouse by its pattern of beams.

Using mercury

Mercury oxidizes in air when it is heated strongly, forming mercuric oxide, which is used in mercury batteries. These are small, powerful batteries with a very long life.

Mercurous chloride is popularly known as calomel. It is normally seen as a white powder. Traditionally calomel dust was used as a teething powder for children because it stops the pain of teething. However, it can also poison children, so it is no longer used.

Calomel has also been used as a protection for seedlings. Before planting, seeds were dusted in the powder; afterwards the calomel was puffed along the rows. This was a good way of preventing club-root pest developing in root crops. However, its use is now restricted because of environmental concerns.

Mercuric chloride is soluble and was, along with arsenic, traditionally used as a poison. In a very dilute form it has been used to help bring on vomiting and as a diuretic.

In modern use it is made into ointments designed to prevent fungal infections (mercurochrome) and as a disinfectant.

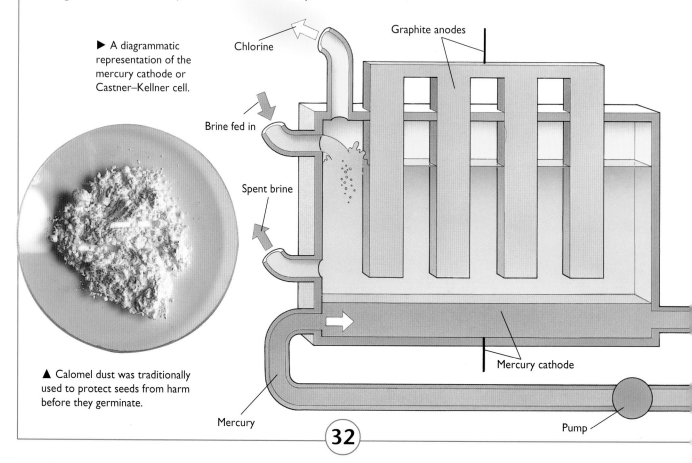

▶ A diagrammatic representation of the mercury cathode or Castner–Kellner cell.

Chlorine

Graphite anodes

Brine fed in

Spent brine

Mercury cathode

▲ Calomel dust was traditionally used to protect seeds from harm before they germinate.

Mercury

Pump

The mercury cathode cell for making chlorine and sodium hydroxide

Chlorine is one of the main chemicals used in modern plastics manufacture. The mercury cathode, also known as the Castner–Kellner cell after its inventors, is an electrolytic process using an electric current to dissociate sodium chloride (brine). It consists of a cell with a graphite anode and a bed of mercury as the cathode. Sodium chloride solution is used as the electrolyte.

A huge current (about 300,000 amps which is about 6000 times as much as a household might use with all of its appliances switched on!) is passed through the cell. Chlorine is released at the anode, where it is collected.

Mercury is used as the cathode because it readily forms an amalgam with other metals. Sodium is a metal ion and makes an amalgam with the mercury. When the mercury cannot absorb any more sodium, the amalgam is carried away and the sodium extracted and made into sodium hydroxide by reaction with water. The refined mercury is then reused and the sodium hydroxide sold to help pay for the process.

cathode: the positive terminal of a battery or the negative electrode of an electrolysis cell.

▼ Mercury oxide batteries are used in cameras, watches, hearing aids and digital calculators.

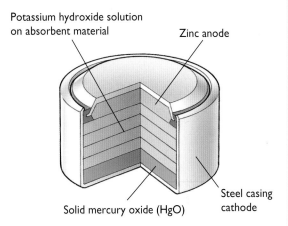

Potassium hydroxide solution on absorbent material

Zinc anode

Solid mercury oxide (HgO)

Steel casing cathode

Mercury dry cells

Mercury dry cells are alkaline cells that use potassium hydroxide as the electrolyte and zinc metal and mercury oxide as the electrodes. During the reaction the mercury oxide is reduced to mercury and the zinc is oxidised (see also page 14).

Mercury cells hold a constant voltage for far longer than normal zinc–carbon dry cells, so small appliances can be maintained at their working voltages for longer. Mercury cells also have a greater power output per unit weight compared with carbon–zinc cells. These characteristics have made them useful for a number of applications, including batteries for hearing aids, although concern about mercury in the environment is making them less widely available.

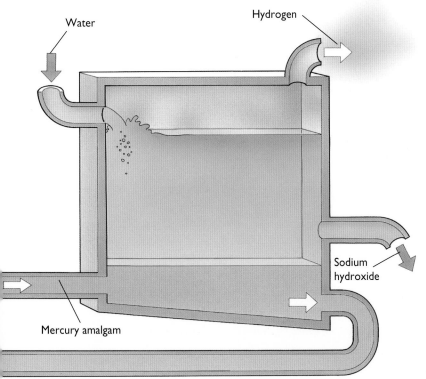

Water

Hydrogen

Sodium hydroxide

Mercury amalgam

Mercury vapour tubes

Discharge tubes are glass tubes filled with a vapour of a metal, often mercury. Electrodes are placed at both ends of the tube. An electricity supply is applied to the electrodes and the vapour ionises and glows.

Mercury vapour lights produce a greenish-blue light. They are widely used for street lighting because they are much more efficient than ordinary (incandescent) light bulbs.

Because the glow from a mercury discharge light is very harsh, the inside of the glass is often coated with a fluorescent material that in turn glows. By choosing the nature of the fluorescent material, a warmer white light can be produced.

Discharge lamps can also be designed to produce brilliant, short duration flashes of light that can be used for night-time beacons for airfields, and for lighthouses and other direction-finding uses.

Sunlamp

Sunlamps use mercury vapour lamps to produce the ultraviolet rays that cause skin to become suntanned.

▶ Mercury vapour lamp

Mercury vapour lamps are made of a glass bulb enclosing a quartz arc tube. Inside the tube are two electrodes, some argon gas and a small amount of mercury. The mercury vaporises when an electric arc forms between the electrodes. (The blue colour is part of the spectrum of light produced; the light also emits strongly in the ultraviolet part of the spectrum, but this is invisible.)

▲ Mercury vapour lighting being used to illuminate a street.

fluorescent: a substance that gives out visible light when struck by invisible waves such as ultraviolet rays.

phosphor: any material that glows when energized by ultraviolet or electron beams such as in fluorescent tubes and cathode ray tubes. Phosphors, such as phosphorus, emit light after the source of excitation is cut off. This is why they glow in the dark. By contrast, fluorescors, such as fluorite, emit light only while they are being excited by ultraviolet light or an electron beam.

vapour: the gaseous form of a substance that is normally a liquid. For example, water vapour is the gaseous form of liquid water.

◀ This experimental design for a mercury vapour lamp is being tested in a laboratory. This lamp uses three times less energy than a conventional filament bulb to produce the same amount of light.

Fluorescent light tubes

These tubes use a mercury arc and a fluorescent coating (phosphor) to produce a very efficient light for use in the home and in offices and shops.

The first person to make a fluorescent tube was Antoine Henri Becquerel. This was in the middle of the 19th century; however, the fluorescent tube has only become widespread since the 1950s.

A fluorescent lamp has two heated coils, one at each end of a tube containing mercury vapour. A mercury arc is formed, but the light emitted is mostly in the invisible, ultraviolet range. To change this into visible light, special phosphors are used as the coating on the inside of the glass. These convert ultraviolet into visible light.

The main advantage of fluorescent tubes is their high efficiency – they consume about one-quarter of the electricity of an incandescent bulb. They also have a long life, often 30–40 times that of an incandescent bulb. However, the starting equipment needed to make the arc ignite adds to the price of the tube fitting.

Amalgams

An amalgam is a mixture, or liquid alloy, of metals with mercury. Amalgams are extremely useful. As early as the 16th century the Venetians used an amalgam of one part tin and two parts mercury as a reflective backing (called "silvering") to make the world's first glass mirrors. (Mirrors are now made with silver and aluminium.)

Amalgams have been used for extracting precious metals such as gold and silver from their ores. They are used widely in industry (see the mercury cathode cell process on page 33), and in dentistry, where the metal fillings are amalgams with silver or gold.

Mercury "eats" holes in thin sheet metal

When mercury is dropped on to the surface of a thin sheet of shiny aluminium, the two elements form an amalgam. Because the sheet is thin, all the aluminium mixes with the mercury at the places where the mercury droplets settle. The result is a pattern of holes where the aluminium has simply become part of the amalgam, which then falls away.

Because it is liquid, mercury cannot develop a coating of unreactive oxide. Thus it is particularly reactive with other metals.

▶ Mercury forms an amalgam with aluminium, corroding holes in this aluminium foil sample.

◄ Small-scale prospectors first remove as much sand and other waste materials as possible, concentrating their ore into the bottom of a pan as shown here. A small amount of mercury is now added and the pan is heated, so that the amalgam can be run off. This picture was taken in the famous Serra Pelada prospecting area of Brazil.

Dental fillings

Dental amalgam is an alloy of 52% mercury, 33% silver, 12.5% tin, 2% copper and 0.5% zinc.

Fillings in the chewing surfaces of teeth are made of wear-resisting amalgams. The silver, tin and mercury are mixed together and forced into the cavity. The mixture then slowly hardens.

Care has to be taken to carve the amalgam into the proper contour shape of the tooth as it is hardening but before it sets hard.

Plastic polymers (which can be white) are usually used in the front of the mouth, where amalgams would look unsightly.

Amalgams in metal refining

Mercury has long been used to extract gold, silver and the other noble metals from the small pieces of ore that are obtained from panning river beds.

The panning process – swirling rock and ore grain around on a shallow plate – cannot completely separate the metal from the rock. If mercury is added to the ore, it dissolves the metal from the rock, forming a paste-like amalgam. The unwanted rock can then be washed away.

To extract the gold or silver from the amalgam, advantage is taken of the low boiling point of mercury. The amalgam is heated over a fire and the mercury vaporises, leaving the pure silver or gold behind. The mercury is collected by condensation and reused.

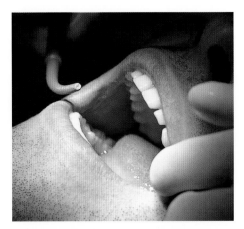

▲ Dental amalgam on the end of a dental instrument about to be pushed into a cavity in a tooth.

Mercury poisoning

Mercury can be a benefit and a hazard. For example, mercurochrome is an organic mercury compound that acts as an antibacterial agent and an effective wound dressing. However, when mercury use is not controlled, disasters often follow.

The hazards of working with mercury have been known for a long time. For example, inhaling mercury vapour can affect the nervous system and in extreme cases can cause severe pneumonia and death. The famous character of the Mad Hatter in Lewis Caroll's *Alice's Adventures in Wonderland* is based on the fact that mercury poisoning was common among the hatters of the 19th century. At this time it was usual to prepare the felt of hats using mercuric nitrate.

Because of such problems, the use of mercury has been drastically reduced. However, some people still use mercury extensively, and in these situations mercury pollution can occur. In 1990 it was feared that gold mining operations in the Brazilian rain forest, where mercury is used to wash gold from sediment, had created pollution in the rivers and was responsible for poisoning fish and humans.

Cosmetics and mercury poisoning

In the past, people placed considerable value on the colours that certain mercury compounds can produce and they used them as cosmetics despite the obvious consequences to their health. By the 16th century, for example, a wide range of cosmetics was available. This included a bright red lip colouring, known as fucus red, which was actually mercuric sulphide. Use on the lips, it was constantly ingested along with food and drink, often leading to fatal consequences.

Another substance, a skin lotion known as Soliman's Water, was used to remove skin blemishes. It was made with sublimate of mercury. If used for long enough, it too could have fatal consequences.

▶ The Mad Hatter (far right) from Lewis Caroll's story *Alice's Adventures in Wonderland.*

▲ Unless great care is taken, mercury used in the gold recovery process (see page 37) can get into the water used for cleaning the gold and cause mercury pollution in nearby rivers.

Methylmercury poisoning

Methylmercury is an organic compound of mercury that is a very effective fungicide. It was first used in the 1940s as a treatment to prevent cereal seeds from rotting before they could germinate. However, the methylmercury remained on the seeds and was eaten by birds, accumulating in their livers and causing damage to their nervous systems. The birds, in turn, were eaten by animals farther up the food chain, concentrating the mercury even more. If such animals are eaten by people, there is a risk of mercury poisoning and thus a threat to human health.

One of the more tragic examples of the careless use of mercury was the disaster at Minimata Bay in Japan in the 1950s. As in many other parts of the world, mercury salts were discharged into the sea along with other effluent from chemical factories. They settled to the bottom of the bay and were thought to be harmless. However, in the sea they were converted to poisonous methylmercury by bacteria. The methylmercury was then eaten by bottom-feeding animals (who dredge the muddy sediments), which, in turn, were eaten by fish. When the fish were caught and sold at local markets, the mercury finally reached people. It had worked its way right through the food chain. The result was a large number of children born with physical and mental abnormalities.

Cadmium

Cadmium is a soft silvery-white metal, much like zinc. The name cadmium comes from the Latin name *cadmia fornacum*. This means "furnace zinc" because it was first discovered as a part of zinc ore.

Cadmium is almost always found associated with zinc and is refined along with zinc from zinc ore. The ore sphalerite contains about one-fifth of one per cent of cadmium.

Cadmium finds widespread use as a catalyst, and its salts are used as colouring pigments in the ceramics industry: cadmium sulphides are yellow.

Cadmium reacts with acids to produce soluble cadmium salts.

Cadmium-plating

Cadmium is a reactive metal that can be used in the same way as zinc to plate steel. A rare metal like cadmium is used in place of zinc because, unlike zinc, it will not be attacked by alkalis such as caustic soda. Also, cadmium can be soldered; zinc cannot.

Cadmium is much more expensive than zinc, so the articles to be plated with cadmium are usually restricted to those that will be in corrosive environments, such as chemical works. Nuts and bolts that are very liable to seize up tight unless they are galvanised are often plated with cadmium.

Cadmium has a low coefficient of friction, so it is also used to coat some bearings.

Test for cadmium compounds

If hydrogen sulphide is bubbled through a solution of cadmium salts, bright yellow cadmium sulphide is precipitated. This is used as a test for cadmium.

▶ Hydrogen sulphide solution is added to cadmium sulphate, resulting in a yellow precipitate of cadmium sulphide.

EQUATION: Testing for cadmium with hydrogen sulphide

Hydrogen sulphide + cadmium sulphate ⇨ cadmium sulphide + sulphuric acid

$$H_2S(g) \quad + \quad CdSO_4(aq) \quad ⇨ \quad CdS(s) \quad + \quad H_2SO_4(l)$$

Uses of cadmium's low melting point

Cadmium melts at 320°C, a very low melting point for a metal. This property is used in a number of ways.

A special kind of solder containing cadmium is used to join pieces of aluminium. Like all solders, aluminium solder must be an alloy of metals that have a low melting point. Aluminium solder is made with 40% cadmium, 50% lead and 10% tin.

Alloys containing cadmium are also used in fire extinguisher systems. Such alloys are combinations of low melting point metals such as cadmium, lead and bismuth. These alloys have the special property that the melting point of the alloy is much lower than the melting points of any of the metals used

One important application of an alloy using cadmium is in the fire-detector nozzles of automatic sprinkler systems used in places like hotels and shops. The nozzles are sealed with an alloy containing cadmium that melts at 70°C. The heat from a fire will raise the temperature of the air near the ceiling, causing the alloy to melt and allowing the water in the pipes to be sprinkled automatically over the fire.

▲ Cadmium is a dense metal that is used to shield nuclear reactors and as the moderating material for the control rods inside the reactor core.

precipitate: tiny solid particles formed as a result of a chemical reaction between two liquids or gases.

solution: a mixture of a liquid and at least one other substance (e.g. salt water). Mixtures can be separated out by physical means, for example by evaporation and cooling.

Separating cadmium and zinc salts

Zinc and cadmium are similar in their behaviour, so it can be difficult to separate them. One way of doing this is to add sodium hydroxide to the mixture of cadmium and zinc salt solution. Hydroxides of zinc and cadmium are formed, but whereas zinc hydroxide is soluble in excess alkali, cadmium hydroxide is insoluble and so forms a precipitate.

Cadmium in batteries

Often it is more convenient to recharge a battery rather than to throw it away. Cadmium is used with nickel to make leakproof and explosion-proof rechargeable batteries.

Rechargeable batteries are not widely used because they cannot supply the current that can be achieved from a disposable primary (dry) cell, and the charge quickly leaks away, meaning that the cells have to be recharged quite often, even if they have not been used.

The most common replacement for zinc dry cells uses nickel and cadmium. The nickel–cadmium cell produces 1.2 volts, slightly lower than the 1.5 volts of a carbon–zinc primary cell. This can mean that, unless the appliance is specifically designed for a nickel–cadmium cell, the performance of the appliance might be slightly below that obtained by using normal dry cells. Thus the brightness of a flashlight will be less using rechargeable batteries than ordinary zinc–carbon cells because the bulb is designed for cells of 1.5 volts. On the other hand, motors in portable screwdrivers are designed for nickel–cadmium batteries and so suffer no loss of performance.

Cadmium and light sensitivity

Cadmium compounds exhibit a range of properties when light shines on them. One of the most commonly used properties is the change in conductivity that occurs when light falls on cadmium sulphide. Cadmium borate is sensitive to ultraviolet light and is used as phosphor material on TV tubes. Cadmium sulphide can also be used as a material that will generate electricity – a solar cell. Most exposure meters in use are photoelectric instruments having either cadmium sulphide or selenium photoelectric cells as light-sensitive elements.

Photovoltaic cells

A device that generates electricity when light falls on it is called a photovoltaic or solar cell.

A range of materials can be used for photovoltaic cells. Usually, two materials are put in contact, the most common being silicon, gallium arsenide, cadmium sulphide and cadmium telluride.

Solar cells made of cadmium sulphide can be made thinner and lighter than silicon cells, but they have a maximum efficiency of about 6%, as opposed to over 14% for silicon cells.

▼ Cadmium compounds are among a range of substances that fluoresce under ultraviolet light. These substances are called fluorescent phosphors, and they make the light-emitting coatings on cathode ray tubes and fluorescent tubes.

Photoelectric devices

A photoelectric device consists of a small piece of photosensitive material, such as cadmium sulphide, connected to a battery.

When light shines on cadmium sulphide, its electrical conductivity increases sharply. This allows it to be used as an automatic switch in such situations as dusk-to-dawn street and home lighting and as the device in automatic cameras.

Cadmium sulphide becomes a conductor when light reaches it because some of the light particles (photons) can make the electrons energetic enough to be free to flow, almost as though they were in a metal.

Cadmium sulphide is especially sensitive to changes in visible light.

phosphor: any material that glows when energized by ultraviolet or electron beams such as in fluorescent tubes and cathode ray tubes. Phosphors, such as phosphorus, emit light after the source of excitation is cut off. This is why they glow in the dark. By contrast, fluorescors, such as fluorite, emit light only while they are being excited by ultraviolet light or an electron beam.

photon: a parcel of light energy.

semiconductor: a material of intermediate conductivity. Semiconductor devices often use silicon when they are made as part of diodes, transistors or integrated circuits.

How an automatic exposure system works

This is a diagram of a circuit used in an automatic camera. The iris in the camera determines the amount of light that reaches the film. A number of small fragments of cadmium sulphide are placed in the path of the light to obtain an average value of the light entering the camera. The light causes a change in the conductivity (resistance) of the cadmium sulphide.

The cadmium sulphide devices are used as though they were variable resistors. A small electronic circuit reads the values of the cadmium sulphide "resistors" and causes the iris to be adjusted until the amount of light reaching the film is appropriate for the correct exposure.

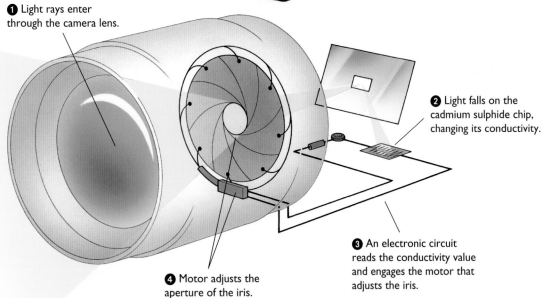

❶ Light rays enter through the camera lens.

❷ Light falls on the cadmium sulphide chip, changing its conductivity.

❸ An electronic circuit reads the conductivity value and engages the motor that adjusts the iris.

❹ Motor adjusts the aperture of the iris.

Key facts about...

Zinc

A soft, blue–white metal, chemical symbol Zn

Required as a trace element by living things

Melts at a lower temperature than most other metals (419°C)

A very reactive metal used to protect other metals from corrosion

Good conductor of electricity

Good conductor of heat

Has no taste

About seven times as dense as water

Atomic number 30, atomic weight about 65

Mercury

A silvery, liquid metal, chemical symbol Hg

Very dense metal (13.6 times as dense as water)

Good conductor of electricity

Easily flows to other shapes

Good conductor of heat

Has no taste

Only 83 parts per billion of the Earth's crust

Atomic number 80, atomic weight about 201

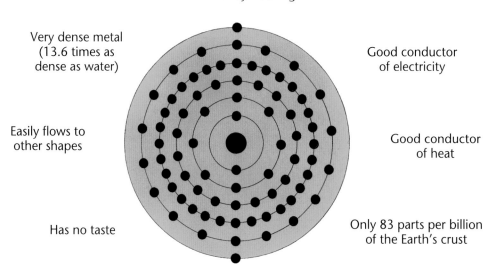

Cadmium

A soft, blue–white metal, chemical symbol Cd

Good conductor of heat

Has no taste

A very reactive metal used as a protective coating on steel

Melts at a lower temperature than most other metals (321°C)

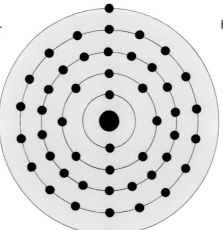

Density 8.65 g/cc

Atomic number 48, atomic weight about 112

SHELL DIAGRAMS

The shell diagrams on these two pages represent an atom of each element. The total number of electrons is shown in the relevant orbitals, or shells, around the central nucleus.

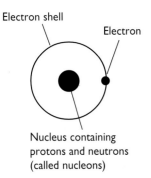

Electron shell

Electron

Nucleus containing protons and neutrons (called nucleons)

▼ When a solution of yellow potassium chromate is dropped into a colourless solution of mercuric nitrate, an orange precipitate of mercuric chromate forms.

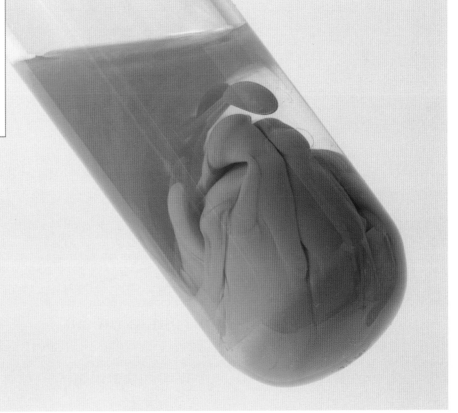

The Periodic Table

The Periodic Table sets out the relationships among the elements of the Universe. According to the Periodic Table, certain elements fall into groups. The pattern of these groups has, in the past, allowed scientists to predict elements that had not at that time been discovered. It can still be used today to predict the properties of unfamiliar elements.

The Periodic Table was first described by a Russian teacher, Dmitry Ivanovich Mendeleev, between 1869 and 1870. He was interested in writing a chemistry textbook, and wanted to show his students that there were certain patterns in the elements that had been discovered. So he set out the elements (of which there were 57 at the time) according to their known properties. On the assumption that there was pattern to the elements, he left blank spaces where elements seemed to be missing. Using this first version of the Periodic Table, he was able to predict in detail the chemical and physical properties of elements that had not yet been discovered. Other scientists began to look for the missing elements, and they soon found them.

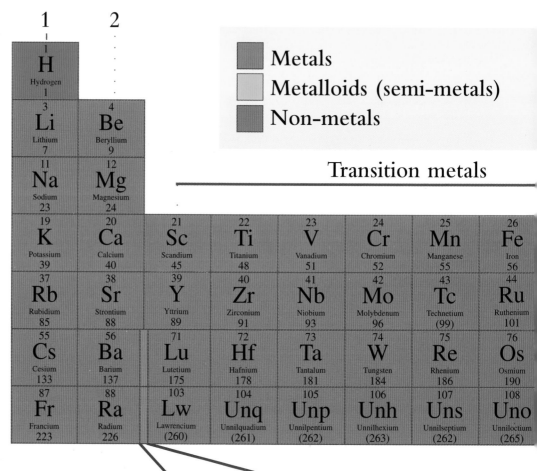

GROUP

Metals
Metalloids (semi-metals)
Non-metals

Transition metals

1	2						
1 **H** Hydrogen 1							
3 **Li** Lithium 7	4 **Be** Beryllium 9						
11 **Na** Sodium 23	12 **Mg** Magnesium 24						
19 **K** Potassium 39	20 **Ca** Calcium 40	21 **Sc** Scandium 45	22 **Ti** Titanium 48	23 **V** Vanadium 51	24 **Cr** Chromium 52	25 **Mn** Manganese 55	26 **Fe** Iron 56
37 **Rb** Rubidium 85	38 **Sr** Strontium 88	39 **Y** Yttrium 89	40 **Zr** Zirconium 91	41 **Nb** Niobium 93	42 **Mo** Molybdenum 96	43 **Tc** Technetium (99)	44 **Ru** Ruthenium 101
55 **Cs** Cesium 133	56 **Ba** Barium 137	71 **Lu** Lutetium 175	72 **Hf** Hafnium 178	73 **Ta** Tantalum 181	74 **W** Tungsten 184	75 **Re** Rhenium 186	76 **Os** Osmium 190
87 **Fr** Francium 223	88 **Ra** Radium 226	103 **Lw** Lawrencium (260)	104 **Unq** Unnilquadium (261)	105 **Unp** Unnilpentium (262)	106 **Unh** Unnilhexium (263)	107 **Uns** Unnilseptium (262)	108 **Uno** Unniloctium (265)

Lanthanide metals

57 **La** Lanthanum 139	58 **Ce** Cerium 140	59 **Pr** Praseodymium 141	60 **Nd** Neodymium 144
89 **Ac** Actinium (227)	90 **Th** Thorium 232	91 **Pa** Protactinium 231	92 **U** Uranium 238

Actinoid metals

Hydrogen did not seem to fit into the table, so he placed it in a box on its own. Otherwise the elements were all placed horizontally. When an element was reached with properties similar to the first one in the top row, a second row was started. By following this rule, similarities among the elements can be found by reading up and down. By reading across the rows, the elements progressively increase their atomic number. This number indicates the number of positively charged particles (protons) in the nucleus of each atom. This is also the number of negatively charged particles (electrons) in the atom.

The chemical properties of an element depend on the number of electrons in the outermost shell.

Atoms can form compounds by sharing electrons in their outermost shells. This explains why atoms with a full set of electrons (like helium, an inert gas) are unreactive, whereas atoms with an incomplete electron shell (such as chlorine) are very reactive. Elements can also combine by the complete transfer of electrons from metals to non-metals and the compounds formed contain ions.

Radioactive elements lose particles from their nucleus and electrons from their surrounding shells. As a result their atomic number changes and they become new elements.

Key:

- Atomic (proton) number: 13
- Symbol: Al
- Name: Aluminium
- Approximate relative atomic mass (Approximate atomic weight): 27

3	4	5	6	7	0
					2 **He** Helium 4
5 **B** Boron 11	6 **C** Carbon 12	7 **N** Nitrogen 14	8 **O** Oxygen 16	9 **F** Fluorine 19	10 **Ne** Neon 20
13 **Al** Aluminium 27	14 **Si** Silicon 28	15 **P** Phosphorus 31	16 **S** Sulphur 32	17 **Cl** Chlorine 35	18 **Ar** Argon 40

27 **Co** Cobalt 59	28 **Ni** Nickel 59	29 **Cu** Copper 64	30 **Zn** Zinc 65	31 **Ga** Gallium 70	32 **Ge** Germanium 73	33 **As** Arsenic 75	34 **Se** Selenium 79	35 **Br** Bromine 80	36 **Kr** Krypton 84
45 **Rh** Rhodium 103	46 **Pd** Palladium 106	47 **Ag** Silver 108	48 **Cd** Cadmium 112	49 **In** Indium 115	50 **Sn** Tin 119	51 **Sb** Antimony 122	52 **Te** Tellurium 128	53 **I** Iodine 127	54 **Xe** Xenon 131
77 **Ir** Iridium 192	78 **Pt** Platinum 195	79 **Au** Gold 197	80 **Hg** Mercury 201	81 **Tl** Thallium 204	82 **Pb** Lead 207	83 **Bi** Bismuth 209	84 **Po** Polonium (209)	85 **At** Astatine (210)	86 **Rn** Radon (222)
109 **Une** Unnilennium (266)									

61 **Pm** Promethium (145)	62 **Sm** Samarium 150	63 **Eu** Europium 152	64 **Gd** Gadolinium 157	65 **Tb** Terbium 159	66 **Dy** Dysprosium 163	67 **Ho** Holmium 165	68 **Er** Erbium 167	69 **Tm** Thulium 169	70 **Yb** Ytterbium 173
93 **Np** Neptunium (237)	94 **Pu** Plutonium (244)	95 **Am** Americium (243)	96 **Cm** Curium (247)	97 **Bk** Berkelium (247)	98 **Cf** Californium (251)	99 **Es** Einsteinium (252)	100 **Fm** Fermium (257)	101 **Md** Mendelevium (258)	102 **No** Nobelium (259)

Understanding equations

As you read through this book, you will notice that many pages contain equations using symbols. If you are not familiar with these symbols, read this page. Symbols make it easy for chemists to write out the reactions that are occurring in a way that allows a better understanding of the processes involved.

Symbols for the elements

The basis of the modern use of symbols for elements dates back to the 19th century. At this time a shorthand was developed using the first letter of the element wherever possible. Thus "O" stands for oxygen, "H" stands for hydrogen

and so on. However, if we were to use only the first letter, then there could be some confusion. For example, nitrogen and nickel would both use the symbols N. To overcome this problem, many elements are symbolised using the first two letters of their full name, and the second letter is lowercase. Thus although nitrogen is N, nickel becomes Ni. Not all symbols come from the English name; many use the Latin name instead. This is why, for example, gold is not G but Au (for the Latin *aurum*) and sodium has the symbol Na, from the Latin *natrium*.

Compounds of elements are made by combining letters. Thus the molecule carbon

Written and symbolic equations

In this book, important chemical equations are briefly stated in words (these are called word equations), and are then shown in their symbolic form along with the states.

What reaction the equation illustrates

EQUATION: The formation of calcium hydroxide

Word equation

Calcium oxide + water ⇨ calcium hydroxide

Symbol equation

$CaO(s)$ + $H_2O(l)$ ⇨ $Ca(OH)_2(aq)$

heated

Sometimes you will find additional descriptions below the symbolic equation.

Symbol showing the state:
s is for solid, *l* is for liquid,
g is for gas and *aq* is for aqueous.

Diagrams

Some of the equations are shown as graphic representations.

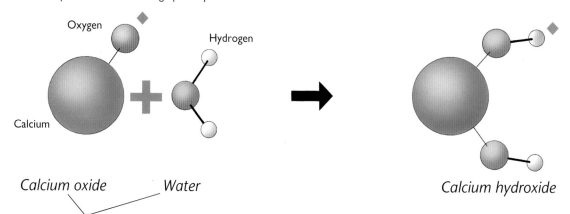

Oxygen

Hydrogen

Calcium

Calcium oxide   *Water*

Calcium hydroxide

Sometimes the written equation is broken up and put below the relevant stages in the graphic representation.

monoxide is CO. By using lowercase letters for the second letter of an element, it is possible to show that cobalt, symbol Co, is not the same as the molecule carbon monoxide, CO.

However, the letters can be made to do much more than this. In many molecules, atoms combine in unequal numbers. So, for example, carbon dioxide has one atom of carbon for every two of oxygen. This is shown by using the number 2 beside the oxygen, and the symbol becomes CO_2.

In practice, some groups of atoms combine as a unit with other substances. Thus, for example, calcium bicarbonate (one of the compounds used in some antacid pills) is written $Ca(HCO_3)_2$. This shows that the part of the substance inside the brackets reacts as a unit and the "2" outside the brackets shows the presence of two such units.

Some substances attract water molecules to themselves. To show this a dot is used. Thus the blue form of copper sulphate is written $CuSO_4.5H_2O$. In this case five molecules of water attract to one of copper sulphate.

When you see the dot, you know that this water can be driven off by heating; it is part of the crystal structure.

In a reaction substances change by rearranging the combinations of atoms. The way they change is shown by using the chemical symbols, placing those that will react (the starting materials, or reactants) on the left and the products of the reaction on the right. Between the two, chemists use an arrow to show which way the reaction is occurring.

It is possible to describe a reaction in words. This gives word equations, which are given throughout this book. However, it is easier to understand what is happening by using an equation containing symbols. These are also given in many places. They are not given when the equations are very complex.

In any equation both sides balance; that is, there must be an equal number of like atoms on both sides of the arrow. When you try to write down reactions, you, too, must balance your equation; you cannot have a few atoms left over at the end!

The symbols in brackets are abbreviations for the physical state of each substance taking part, so that (*s*) is used for solid, (*l*) for liquid, (*g*) for gas and (*aq*) for an aqueous solution, that is, a solution of a substance dissolved in water.

Atoms and ions
Each sphere represents a particle of an element. A particle can be an atom or an ion. Each atom or ion is associated with other atoms or ions through bonds – forces of attraction. The size of the particles and the nature of the bonds can be extremely important in determining the nature of the reaction or the properties of the compound.

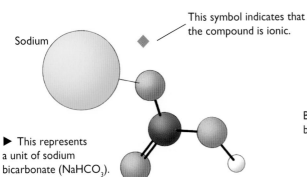

Sodium

This symbol indicates that the compound is ionic.

▶ This represents a unit of sodium bicarbonate ($NaHCO_3$).

The term "unit" is sometimes used to simplify the representation of a combination of ions.

Chemical symbols, equations and diagrams
The arrangement of any molecule or compound can be shown in one of the two ways shown below, depending on which gives the clearer picture. The left-hand diagram is called a ball-and-stick diagram because it uses rods and spheres to show the structure of the material. This example shows water, H_2O. There are two hydrogen atoms and one oxygen atom.

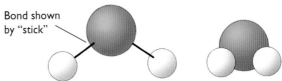

Bond shown by "stick"

Colours too
The colours of each of the particles help differentiate the elements involved. The diagram can then be matched to the written and symbolic equation given with the diagram. In the case above, oxygen is red and hydrogen is grey.

Glossary of technical terms

absorb: to soak up a substance. Compare to adsorb.

acetone: a petroleum-based solvent.

acid: compounds containing hydrogen which can attack and dissolve many substances. Acids are described as weak or strong, dilute or concentrated, mineral or organic.

acidity: a general term for the strength of an acid in a solution.

acid rain: rain that is contaminated by acid gases such as sulphur dioxide and nitrogen oxides released by pollution.

adsorb/adsorption: to "collect" gas molecules or other particles on to the *surface* of a substance. They are not chemically combined and can be removed. (The process is called "adsorption".) Compare to absorb.

alchemy: the traditional "art" of working with chemicals that prevailed through the Middle Ages. One of the main challenges of alchemy was to make gold from lead. Alchemy faded away as scientific chemistry was developed in the 17th century.

alkali: a base in solution.

alkaline: the opposite of acidic. Alkalis are bases that dissolve, and alkaline materials are called basic materials. Solutions of alkalis have a pH greater than 7.0 because they contain relatively few hydrogen ions.

alloy: a mixture of a metal and various other elements.

alpha particle: a stable combination of two protons and two neutrons, which is ejected from the nucleus of a radioactive atom as it decays. An alpha particle is also the nucleus of the atom of helium. If it captures two electrons it can become a neutral helium atom.

amalgam: a liquid alloy of mercury with another metal.

amino acid: amino acids are organic compounds that are the building blocks for the proteins in the body.

amorphous: a solid in which the atoms are not arranged regularly (i.e. "glassy"). Compare with crystalline.

amphoteric: a metal that will react with both acids and alkalis.

anhydrous: a substance from which water has been removed by heating. Many hydrated salts are crystalline. When they are heated and the water is driven off, the material changes to an anhydrous powder.

anion: a negatively charged atom or group of atoms.

anode: the negative terminal of a battery or the positive electrode of an electrolysis cell.

anodising: a process that uses the effect of electrolysis to make a surface corrosion-resistant.

antacid: a common name for any compound that reacts with stomach acid to neutralise it.

antioxidant: a substance that prevents oxidation of some other substance.

aqueous: a solid dissolved in water. Usually used as "aqueous solution".

atom: the smallest particle of an element.

atomic number: the number of electrons or the number of protons in an atom.

atomised: broken up into a very fine mist. The term is used in connection with sprays and engine fuel systems.

aurora: the "northern lights" and "southern lights" that show as coloured bands of light in the night sky at high latitudes. They are associated with the way cosmic rays interact with oxygen and nitrogen in the air.

basalt: an igneous rock with a low proportion of silica (usually below 55%). It has microscopically small crystals.

base: a compound that may be soapy to the touch and that can react with an acid in water to form a salt and water.

battery: a series of electrochemical cells.

bauxite: an ore of aluminium, of which about half is aluminium oxide.

becquerel: a unit of radiation equal to one nuclear disintegration per second.

beta particle: a form of radiation in which electrons are emitted from an atom as the nucleus breaks down.

bleach: a substance that removes stains from materials either by oxidising or reducing the staining compound.

boiling point: the temperature at which a liquid boils, changing from a liquid to a gas.

bond: chemical bonding is either a transfer or sharing of electrons by two or more atoms. There are a number of types of chemical bond, some very strong (such as covalent bonds), others weak (such as hydrogen bonds). Chemical bonds form because the linked molecule is more stable than the unlinked atoms from which it formed. For example, the hydrogen molecule (H_2) is more stable than single atoms of hydrogen, which is why hydrogen gas is always found as molecules of two hydrogen atoms.

brass: a metal alloy principally of copper and zinc.

brazing: a form of soldering, in which brass is used as the joining metal.

brine: a solution of salt (sodium chloride) in water.

bronze: an alloy principally of copper and tin.

buffer: a chemistry term meaning a mixture of substances in solution that resists a change in the acidity or alkalinity of the solution.

capillary action: the tendency of a liquid to be sucked into small spaces, such as between objects and through narrow-pore tubes. The force to do this comes from surface tension.

catalyst: a substance that speeds up a chemical reaction but itself remains unaltered at the end of the reaction.

cathode: the positive terminal of a battery or the negative electrode of an electrolysis cell.

cathodic protection: the technique of making the object that is to be protected from corrosion into the cathode of a cell. For example, a material, such as steel, is protected by coupling it with a more reactive metal, such as magnesium. Steel forms the cathode and magnesium the anode. Zinc protects steel in the same way.

cation: a positively charged atom or group of atoms.

caustic: a substance that can cause burns if it touches the skin.

cell: a vessel containing two electrodes and an electrolyte that can act as an electrical conductor.

ceramic: a material based on clay minerals, which has been heated so that it has chemically hardened.

chalk: a pure form of calcium carbonate made of the crushed bodies of microscopic sea creatures, such as plankton and algae.

change of state: a change between one of the three states of matter, solid, liquid and gas.

chlorination: adding chlorine to a substance.

cladding: a surface sheet of material designed to protect other materials from corrosion.

clay: a microscopically small plate-like mineral that makes up the bulk of many soils. It has a sticky feel when wet.

combustion: the special case of oxidisation of a substance where a considerable amount of heat and usually light are given out. Combustion is often referred to as "burning".

compound: a chemical consisting of two or more elements chemically bonded together. Calcium atoms can combine with carbon atoms and oxygen atoms to make calcium carbonate, a compound of all three atoms.

condensation nuclei: microscopic particles of dust, salt and other materials suspended in the air, which attract water molecules.

conduction: (i) the exchange of heat (heat conduction) by contact with another object or (ii) allowing the flow of electrons (electrical conduction).

convection: the exchange of heat energy with the surroundings produced by the flow of a fluid due to being heated or cooled.

corrosion: the *slow* decay of a substance resulting from contact with gases and liquids in the environment. The term is often applied to metals. Rust is the corrosion of iron.

corrosive: a substance, either an acid or an alkali, that *rapidly* attacks a wide range of other substances.

cosmic rays: particles that fly through space and bombard all atoms on the Earth's surface. When they interact with the atmosphere they produce showers of secondary particles.

covalent bond: the most common form of strong chemical bonding, which occurs when two atoms *share* electrons.

cracking: breaking down complex molecules into simpler components. It is a term particularly used in oil refining.

crude oil: a chemical mixture of petroleum liquids. Crude oil forms the raw material for an oil refinery.

crystal: a substance that has grown freely so that it can develop external faces. Compare with crystalline, where the atoms are not free to form individual crystals and amorphous where the atoms are arranged irregularly.

crystalline: the organisation of atoms into a rigid "honeycomb-like" pattern without distinct crystal faces.

crystal systems: seven patterns or systems into which all of the world's crystals can be grouped. They are: cubic, hexagonal, rhombohedral, tetragonal, orthorhombic, monoclinic and triclinic.

cubic crystal system: groupings of crystals that look like cubes.

curie: a unit of radiation. The amount of radiation emitted by 1 g of radium each second. (The curie is equal to 37 billion becquerels.)

current: an electric current is produced by a flow of electrons through a conducting solid or ions through a conducting liquid.

decay (radioactive decay): the way that a radioactive element changes into another element because of loss of mass through radiation. For example uranium decays (changes) to lead.

decompose: to break down a substance (for example by heat or with the aid of a catalyst) into simpler components. In such a chemical reaction only one substance is involved.

dehydration: the removal of water from a substance by heating it, placing it in a dry atmosphere, or through the action of a drying agent.

density: the mass per unit volume (e.g. g/cc)

desertification: a process whereby a soil is allowed to become degraded to a state in which crops can no longer grow, i.e. desert-like. Chemical desertification is usually the result of contamination with halides because of poor irrigation practices.

detergent: a petroleum-based chemical that removes dirt.

diaphragm: a semipermeable membrane – a kind of ultra-fine mesh filter – that will allow only small ions to pass through. It is used in the electrolysis of brine.

diffusion: the slow mixing of one substance with another until the two substances are evenly mixed.

digestive tract: the system of the body that forms the pathway for food and its waste products. It begins at the mouth and includes the stomach and the intestines.

dilute acid: an acid whose concentration has been reduced by a large proportion of water.

diode: a semiconducting device that allows an electric current to flow in only one direction.

disinfectant: a chemical that kills bacteria and other microorganisms.

dissociate: to break apart. In the case of acids it means to break up forming hydrogen ions. This is an example of ionisation. Strong acids dissociate completely. Weak acids are not completely ionised and a solution of a weak acid has a relatively low concentration of hydrogen ions.

dissolve: to break down a substance in a solution without a resultant reaction.

distillation: the process of separating mixtures by condensing the vapours through cooling.

doping: adding metal atoms to a region of silicon to make it semiconducting.

dye: a coloured substance that will stick to another substance, so that both appear coloured.

electrode: a conductor that forms one terminal of a cell.

electrolysis: an electrical–chemical process that uses an electric current to cause the break up of a compound and the movement of metal ions in a solution. The process happens in many natural situations (as for example in rusting) and is also commonly used in industry for purifying (refining) metals or for plating metal objects with a fine, even metal coating.

electrolyte: a solution that conducts electricity.

electron: a tiny, negatively charged particle that is part of an atom. The flow of electrons through a solid material such as a wire produces an electric current.

electroplating: depositing a thin layer of a metal onto the surface of another substance using electrolysis.

element: a substance that cannot be decomposed into simpler substances by chemical means

emulsion: tiny droplets of one substance dispersed in another. A common oil in water emulsion is milk. The tiny droplets in an emulsion tend to come together, so another stabilising substance is often needed to wrap the particles of grease and oil in a stable coat. Soaps and detergents are such agents. Photographic film is an example of a solid emulsion.

endothermic reaction: a reaction that takes heat from the surroundings. The reaction of carbon monoxide with a metal oxide is an example.

enzyme: organic catalysts in the form of proteins in the body that speed up chemical reactions. Every living cell contains hundreds of enzymes, which ensure that the processes of life continue. Should enzymes be made inoperative, such as through mercury poisoning, then death follows.

ester: organic compounds, formed by the reaction of an alcohol with an acid, which often have a fruity taste.

evaporation: the change of state of a liquid to a gas. Evaporation happens below the boiling point and is used as a method of separating out the materials in a solution.

exothermic reaction: a reaction that gives heat to the surroundings. Many oxidation reactions, for example, give out heat.

explosive: a substance which, when a shock is applied to it, decomposes very rapidly, releasing a very large amount of heat and creating a large volume of gases as a shock wave.

extrusion: forming a shape by pushing it through a die. For example, toothpaste is extruded through the cap (die) of the toothpaste tube.

fallout: radioactive particles that reach the ground from radioactive materials in the atmosphere.

fat: semi-solid energy-rich compounds derived from plants or animals and which are made of carbon, hydrogen and oxygen. Scientists call these esters.

feldspar: a mineral consisting of sheets of aluminium silicate. This is the mineral from which the clay in soils is made.

fertile: able to provide the nutrients needed for unrestricted plant growth.

filtration: the separation of a liquid from a solid using a membrane with small holes.

fission: the breakdown of the structure of an atom, popularly called "splitting the atom" because the atom is split into approximately two other nuclei. This is different from, for example, the small change that happens when radioactivity is emitted.

fixation of nitrogen: the processes that natural organisms, such as bacteria, use to turn the nitrogen of the air into ammonium compounds.

fixing: making solid and liquid nitrogen-containing compounds from nitrogen gas. The compounds that are formed can be used as fertilisers.

fluid: able to flow; either a liquid or a gas.

fluorescent: a substance that gives out visible light when struck by invisible waves such as ultraviolet rays.

flux: a material used to make it easier for a liquid to flow. A flux dissolves metal oxides and so prevents a metal from oxidising while being heated.

foam: a substance that is sufficiently gelatinous to be able to contain bubbles of gas. The gas bulks up the substance, making it behave as though it were semi-rigid.

fossil fuels: hydrocarbon compounds that have been formed from buried plant and animal remains. High pressures and temperatures lasting over millions of years are required. The fossil fuels are coal, oil and natural gas.

fraction: a group of similar components of a mixture. In the petroleum industry the light fractions of crude oil are those with the smallest molecules, while the medium and heavy fractions have larger molecules.

free radical: a very reactive atom or group with a "spare" electron.

freezing point: the temperature at which a substance changes from a liquid to a solid. It is the same temperature as the melting point.

fuel: a concentrated form of chemical energy. The main sources of fuels (called fossil fuels because they were formed by geological processes) are coal, crude oil and natural gas. Products include methane, propane and gasoline. The fuel for stars and space vehicles is hydrogen.

fuel rods: rods of uranium or other radioactive material used as a fuel in nuclear power stations.

fuming: an unstable liquid that gives off a gas. Very concentrated acid solutions are often fuming solutions.

fungicide: any chemical that is designed to kill fungi and control the spread of fungal spores.

fusion: combining atoms to form a heavier atom.

galvanising: applying a thin zinc coating to protect another metal.

gamma rays: waves of radiation produced as the nucleus of a radioactive element rearranges itself into a tighter cluster of protons and neutrons. Gamma rays carry enough energy to damage living cells.

gangue: the unwanted material in an ore.

gas: a form of matter in which the molecules form no definite shape and are free to move about to fill any vessel they are put in.

gelatinous: a term meaning made with water. Because a gelatinous precipitate is mostly water, it is of a similar density to water and will float or lie suspended in the liquid.

gelling agent: a semi-solid jelly-like substance.

gemstone: a wide range of minerals valued by people, both as crystals (such as emerald) and as decorative stones (such as agate). There is no single chemical formula for a gemstone.

glass: a transparent silicate without any crystal growth. It has a glassy lustre and breaks with a curved fracture. Note that some minerals have all these features and are therefore natural glasses. Household glass is a synthetic silicate.

glucose: the most common of the natural sugars. It occurs as the polymer known as cellulose, the fibre in plants. Starch is also a form of glucose. The breakdown of glucose provides the energy that animals need for life.

granite: an igneous rock with a high proportion of silica (usually over 65%). It has well-developed large crystals. The largest pink, grey or white crystals are feldspar.

Greenhouse Effect: an increase of the global air temperature as a result of heat released from burning fossil fuels being absorbed by carbon dioxide in the atmosphere.

gypsum: the name for calcium sulphate. It is commonly found as Plaster of Paris and wallboards.

half-life: the time it takes for the radiation coming from a sample of a radioactive element to decrease by half.

halide: a salt of one of the halogens (fluorine, chlorine, bromine and iodine).

halite: the mineral made of sodium chloride.

halogen: one of a group of elements including chlorine, bromine, iodine and fluorine.

heat-producing: see exothermic reaction.

high explosive: a form of explosive that will only work when it receives a shock from another explosive. High explosives are much more powerful than ordinary explosives. Gunpowder is not a high explosive.

hydrate: a solid compound in crystalline form that contains molecular water. Hydrates commonly form when a solution of a soluble salt is evaporated. The water that forms part of a hydrate crystal is known as the "water of crystallization". It can usually be removed by heating, leaving an anhydrous salt.

hydration: the absorption of water by a substance. Hydrated materials are not "wet" but remain firm, apparently dry, solids. In some cases, hydration makes the substance change colour, in many other cases there is no colour change, simply a change in volume.

hydrocarbon: a compound in which only hydrogen and carbon atoms are present. Most fuels are hydrocarbons, as is the simple plastic polyethene (known as polythene).

hydrogen bond: a type of attractive force that holds one molecule to another. It is one of the weaker forms of intermolecular attractive force.

hydrothermal: a process in which hot water is involved. It is usually used in the context of rock formation because hot water and other fluids sent outwards from liquid magmas are important carriers of metals and the minerals that form gemstones.

igneous rock: a rock that has solidified from molten rock, either volcanic lava on the Earth's surface or magma deep underground. In either case the rock develops a network of interlocking crystals.

incendiary: a substance designed to cause burning.

indicator: a substance or mixture of substances that change colour with acidity or alkalinity.

inert: nonreactive.

infra-red radiation: a form of light radiation where the wavelength of the waves is slightly longer than visible light. Most heat radiation is in the infra-red band.

insoluble: a substance that will not dissolve.

ion: an atom, or group of atoms, that has gained or lost one or more electrons and so developed an electrical charge. Ions behave differently from electrically neutral atoms and molecules. They can move in an electric field, and they can also bind strongly to solvent molecules such as water. Positively charged ions are called cations; negatively charged ions are called anions. Ions carry electrical current through solutions.

ionic bond: the form of bonding that occurs between two ions when the ions have opposite charges. Sodium cations bond with chloride anions to form common salt (NaCl) when a salty solution is evaporated. Ionic bonds are strong bonds except in the presence of a solvent.

ionise: to break up neutral molecules into oppositely charged ions or to convert atoms into ions by the loss of electrons.

ionisation: a process that creates ions.

irrigation: the application of water to fields to help plants grow during times when natural rainfall is sparse.

isotope: atoms that have the same number of protons in their nucleus, but which have different masses; for example, carbon-12 and carbon-14.

latent heat: the amount of heat that is absorbed or released during the process of changing state between gas, liquid or solid. For example, heat is absorbed when a substance melts and it is released again when the substance solidifies.

latex: (the Latin word for "liquid") a suspension of small polymer particles in water. The rubber that flows from a rubber tree is a natural latex. Some synthetic polymers are made as latexes, allowing polymerisation to take place in water.

lava: the material that flows from a volcano.

limestone: a form of calcium carbonate rock that is often formed of lime mud. Most limestones are light grey and have abundant fossils.

liquid: a form of matter that has a fixed volume but no fixed shape.

lode: a deposit in which a number of veins of a metal found close together.

lustre: the shininess of a substance.

magma: the molten rock that forms a balloon-shaped chamber in the rock below a volcano. It is fed by rock moving upwards from below the crust.

marble: a form of limestone that has been "baked" while deep inside mountains. This has caused the limestone to melt and reform into small interlocking crystals, making marble harder than limestone.

mass: the amount of matter in an object. In everyday use, the word weight is often used to mean mass.

melting point: the temperature at which a substance changes state from a solid to a liquid. It is the same as freezing point.

membrane: a thin flexible sheet. A semipermeable membrane has microscopic holes of a size that will selectively allow some ions and molecules to pass through but hold others back. It thus acts as a kind of sieve.

meniscus: the curved surface of a liquid that forms when it rises in a small bore, or capillary tube. The meniscus is convex (bulges upwards) for mercury and is concave (sags downwards) for water.

metal: a substance with a lustre, the ability to conduct heat and electricity and which is not brittle.

metallic bonding: a kind of bonding in which atoms reside in a "sea" of mobile electrons. This type of bonding allows metals to be good conductors and means that they are not brittle

metamorphic rock: formed either from igneous or sedimentary rocks, by heat and or pressure. Metamorphic rocks form deep inside mountains during periods of mountain building. They result from the remelting of rocks during which process crystals are able to grow. Metamorphic rocks often show signs of banding and partial melting.

micronutrient: an element that the body requires in small amounts. Another term is trace element.

mineral: a solid substance made of just one element or chemical compound. Calcite is a mineral because it consists only of calcium carbonate, halite is a mineral because it contains only sodium chloride, quartz is a mineral because it consists of only silicon dioxide.

mineral acid: an acid that does not contain carbon and that attacks minerals. Hydrochloric, sulphuric and nitric acids are the main mineral acids.

mineral-laden: a solution close to saturation.

mixture: a material that can be separated out into two or more substances using physical means.

molecule: a group of two or more atoms held together by chemical bonds.

monoclinic system: a grouping of crystals that look like double-ended chisel blades.

monomer: a building block of a larger chain molecule ("mono" means one, "mer" means part).

mordant: any chemical that allows dyes to stick to other substances.

native metal: a pure form of a metal, not combined as a compound. Native metal is more common in poorly reactive elements than in those that are very reactive.

neutralisation: the reaction of acids and bases to produce a salt and water. The reaction causes hydrogen from the acid and hydroxide from the base to be changed to water. For example, hydrochloric acid reacts with sodium hydroxide to form common salt and water. The term is more generally used for any reaction where the pH changes towards 7.0, which is the pH of a neutral solution.

neutron: a particle inside the nucleus of an atom that is neutral and has no charge.

noncombustible: a substance that will not burn.

noble metal: silver, gold, platinum, and mercury. These are the least reactive metals.

nuclear energy: the heat energy produced as part of the changes that take place in the core, or nucleus, of an element's atoms.

nuclear reactions: reactions that occur in the core, or nucleus of an atom.

nutrients: soluble ions that are essential to life.

octane: one of the substances contained in fuel.

ore: a rock containing enough of a useful substance to make mining it worthwhile.

organic acid: an acid containing carbon and hydrogen.

organic substance: a substance that contains carbon.

osmosis: a process where molecules of a liquid solvent move through a membrane (filter) from a region of low concentration to a region of high concentration of solute.

oxidation: a reaction in which the oxidising agent removes electrons. (Note that oxidising agents do not have to contain oxygen.)

oxide: a compound that includes oxygen and one other element.

oxidise: the process of gaining oxygen. This can be part of a controlled chemical reaction, or it can be the result of exposing a substance to the air, where oxidation (a form of corrosion) will occur slowly, perhaps over months or years.

oxidising agent: a substance that removes electrons from another substance (and therefore is itself reduced).

ozone: a form of oxygen whose molecules contain three atoms of oxygen. Ozone is regarded as a beneficial gas when high in the atmosphere because it blocks ultraviolet rays. It is a harmful gas when breathed in, so low level ozone, which is produced as part of city smog, is regarded as a form of pollution. The ozone layer is the uppermost part of the stratosphere.

pan: the name given to a shallow pond of liquid. Pans are mainly used for separating solutions by evaporation.

patina: a surface coating that develops on metals and protects them from further corrosion.

percolate: to move slowly through the pores of a rock.

period: a row in the Periodic Table.

Periodic Table: a chart organising elements by atomic number and chemical properties into groups and periods.

pesticide: any chemical that is designed to control pests (unwanted organisms) that are harmful to plants or animals.

petroleum: a natural mixture of a range of gases, liquids and solids derived from the decomposed remains of plants and animals.

pH: a measure of the hydrogen ion concentration in a liquid. Neutral is pH 7.0; numbers greater than this are alkaline, smaller numbers are acidic.

phosphor: any material that glows when energized by ultraviolet or electron beams such as in fluorescent tubes and cathode ray tubes. Phosphors, such as phosphorus, emit light after the source of excitation is cut off. This is why they glow in the dark. By contrast, fluorescors, such as fluorite, emit light only while they are being excited by ultraviolet light or an electron beam.

photon: a parcel of light energy.

photosynthesis: the process by which plants use the energy of the Sun to make the compounds they need for life. In photosynthesis, six molecules of carbon dioxide from the air combine with six molecules of water, forming one molecule of glucose (sugar) and releasing six molecules of oxygen back into the atmosphere.

pigment: any solid material used to give a liquid a colour.

placer deposit: a kind of ore body made of a sediment that contains fragments of gold ore eroded from a mother lode and transported by rivers and/or ocean currents.

plastic (material): a carbon-based material consisting of long chains (polymers) of simple molecules. The word plastic is commonly restricted to synthetic polymers.

plastic (property): a material is plastic if it can be made to change shape easily. Plastic materials will remain in the new shape. (Compare with elastic, a property where a material goes back to its original shape.)

plating: adding a thin coat of one material to another to make it resistant to corrosion.

playa: a dried-up lake bed that is covered with salt deposits. From the Spanish word for beach.

poison gas: a form of gas that is used intentionally to produce widespread injury and death. (Many gases are poisonous, which is why many chemical reactions are performed in laboratory fume chambers, but they are a byproduct of a reaction and not intended to cause harm.)

polymer: a compound that is made of long chains by combining molecules (called monomers) as repeating units. ("Poly" means many, "mer" means part).

polymerisation: a chemical reaction in which large numbers of similar molecules arrange themselves into large molecules, usually long chains. This process usually happens when there is a suitable catalyst present. For example, ethene reacts to form polythene in the presence of certain catalysts.

porous: a material containing many small holes or cracks. Quite often the pores are connected, and liquids, such as water or oil, can move through them.

precious metal: silver, gold, platinum, iridium, and palladium. Each is prized for its rarity. This category is the equivalent of precious stones, or gemstones, for minerals.

precipitate: tiny solid particles formed as a result of a chemical reaction between two liquids or gases.

preservative: a substance that prevents the natural organic decay processes from occurring. Many substances can be used safely for this purpose, including sulphites and nitrogen gas.

product: a substance produced by a chemical reaction.

protein: molecules that help to build tissue and bone and therefore make new body cells. Proteins contain amino acids.

proton: a positively charged particle in the nucleus of an atom that balances out the charge of the surrounding electrons

pyrite: "mineral of fire". This name comes from the fact that pyrite (iron sulphide) will give off sparks if struck with a stone.

pyrometallurgy: refining a metal from its ore using heat. A blast furnace or smelter is the main equipment used.

radiation: the exchange of energy with the surroundings through the transmission of waves or particles of energy. Radiation is a form of energy transfer that can happen through space; no intervening medium is required (as would be the case for conduction and convection).

radioactive: a material that emits radiation or particles from the nucleus of its atoms.

radioactive decay: a change in a radioactive element due to loss of mass through radiation. For example uranium decays (changes) to lead.

radioisotope: a shortened version of the phrase radioactive isotope.

radiotracer: a radioactive isotope that is added to a stable, nonradioactive material in order to trace how it moves and its concentration.

reaction: the recombination of two substances using parts of each substance to produce new substances.

reactivity: the tendency of a substance to react with other substances. The term is most widely used in comparing the reactivity of metals. Metals are arranged in a reactivity series.

reagent: a starting material for a reaction.

recycling: the reuse of a material to save the time and energy required to extract new material from the Earth and to conserve non-renewable resources.

redox reaction: a reaction that involves reduction and oxidation.

reducing agent: a substance that gives electrons to another substance. Carbon monoxide is a reducing agent when passed over copper oxide, turning it to copper and producing carbon dioxide gas. Similarly, iron oxide is reduced to iron in a blast furnace. Sulphur dioxide is a reducing agent, used for bleaching bread.

reduction: the removal of oxygen from a substance. See also: oxidation.

refining: separating a mixture into the simpler substances of which it is made. In the case of a rock, it means the extraction of the metal that is mixed up in the rock. In the case of oil it means separating out the fractions of which it is made.

refractive index: the property of a transparent material that controls the angle at which total internal reflection will occur. The greater the refractive index, the more reflective the material will be.

resin: natural or synthetic polymers that can be moulded into solid objects or spun into thread.

rust: the corrosion of iron and steel.

saline: a solution in which most of the dissolved matter is sodium chloride (common salt).

salinisation: the concentration of salts, especially sodium chloride, in the upper layers of a soil due to poor methods of irrigation.

salts: compounds, often involving a metal, that are the reaction products of acids and bases. (Note "salt" is also the common word for sodium chloride, common salt or table salt.)

saponification: the term for a reaction between a fat and a base that produces a soap.

saturated: a state where a liquid can hold no more of a substance. If any more of the substance is added, it will not dissolve.

saturated solution: a solution that holds the maximum possible amount of dissolved material. The amount of material in solution varies with the temperature; cold solutions can hold less dissolved solid material than hot solutions. Gases are more soluble in cold liquids than hot liquids.

sediment: material that settles out at the bottom of a liquid when it is still.

semiconductor: a material of intermediate conductivity. Semiconductor devices often use silicon when they are made as part of diodes, transistors or integrated circuits.

semipermeable membrane: a thin (membrane) of material that acts as a fine sieve, allowing small molecules to pass, but holding large molecules back.

silicate: a compound containing silicon and oxygen (known as silica).

sintering: a process that happens at moderately high temperatures in some compounds. Grains begin to fuse together even through they do not melt. The most widespread example of sintering happens during the firing of clays to make ceramics.

slag: a mixture of substances that are waste products of a furnace. Most slags are composed mainly of silicates.

smelting: roasting a substance in order to extract the metal contained in it.

smog: a mixture of smoke and fog. The term is used to describe city fogs in which there is a large proportion of particulate matter (tiny pieces of carbon from exhausts) and also a high concentration of sulphur and nitrogen gases and probably ozone.

soldering: joining together two pieces of metal using solder, an alloy with a low melting point.

solid: a form of matter where a substance has a definite shape.

soluble: a substance that will readily dissolve in a solvent.

solute: the substance that dissolves in a solution (e.g. sodium chloride in salt water).

solution: a mixture of a liquid and at least one other substance (e.g. salt water). Mixtures can be separated out by physical means, for example by evaporation and cooling.

solvent: the main substance in a solution (e.g. water in salt water).

spontaneous combustion: the effect of a very reactive material beginning to oxidise very quickly and bursting into flame.

stable: able to exist without changing into another substance.

stratosphere: the part of the Earth's atmosphere that lies immediately above the region in which clouds form. It occurs between 12 and 50 km above the Earth's surface.

strong acid: an acid that has completely dissociated (ionised) in water. Mineral acids are strong acids.

sublimation: the change of a substance from solid to gas, or vica versa, without going through a liquid phase.

substance: a type of material, including mixtures.

sulphate: a compound that includes sulphur and oxygen, for example, calcium sulphate or gypsum.

sulphide: a sulphur compound that contains no oxygen.

sulphite: a sulphur compound that contains less oxygen than a sulphate.

surface tension: the force that operates on the surface of a liquid, which makes it act as though it were covered with an invisible elastic film.

suspension: tiny particles suspended in a liquid.

synthetic: does not occur naturally, but has to be manufactured.

tarnish: a coating that develops as a result of the reaction between a metal and substances in the air. The most common form of tarnishing is a very thin transparent oxide coating.

thermonuclear reactions: reactions that occur within atoms due to fusion, releasing an immensely concentrated amount of energy.

thermoplastic: a plastic that will soften, can repeatedly be moulded it into shape on heating and will set into the moulded shape as it cools.

thermoset: a plastic that will set into a moulded shape as it cools, but which cannot be made soft by reheating.

titration: a process of dripping one liquid into another in order to find out the amount needed to cause a neutral solution. An indicator is used to signal change.

toxic: poisonous enough to cause death.

translucent: almost transparent.

transmutation: the change of one element into another.

vapour: the gaseous form of a substance that is normally a liquid. For example, water vapour is the gaseous form of liquid water.

vein: a mineral deposit different from, and usually cutting across, the surrounding rocks. Most mineral and metal-bearing veins are deposits filling fractures. The veins were filled by hot, mineral-rich waters rising upwards from liquid volcanic magma. They are important sources of many metals, such as silver and gold, and also minerals such as gemstones. Veins are usually narrow, and were best suited to hand-mining. They are less exploited in the modern machine age.

viscous: slow moving, syrupy. A liquid that has a low viscosity is said to be mobile.

vitreous: glass-like.

volatile: readily forms a gas.

vulcanisation: forming cross-links between polymer chains to increase the strength of the whole polymer. Rubbers are vulcanised using sulphur when making tyres and other strong materials.

weak acid: an acid that has only partly dissociated (ionised) in water. Most organic acids are weak acids.

weather: a term used by Earth scientists and derived from "weathering", meaning to react with water and gases of the environment.

weathering: the slow natural processes that break down rocks and reduce them to small fragments either by mechanical or chemical means.

welding: fusing two pieces of metal together using heat.

X-rays: a form of very short wave radiation.

Index